Statistics Safari:

A 10-day introductory guided tour for anyone
who loves people more than numbers

by Dr. Ken Tangen

Saddleback Press
www.saddlebackpress.com

Cover design & illustrations at the beginning of each chapter are by Andrew Bosley. All other illustrations were created by the author who notes: "Andrew drew all of the great looking illustrations. I created the rest and they are clear evidence of why I went into psychology and not art."

Special thanks to Corianne Rogalsky and Denise Robb.

ISBN: 0-9765360-2-1

Printed in the United States of America

 Statistics shouldn't be hard or scary. Although it might not be as fun as a vacation, it shouldn't be as hard as climbing a mountain. It should be more like a guided tour.

I've written this to be more of a guide book than a textbook. I've tried to pick out interesting sites, tell stories about them, and generally point you in the right direction. Naturally, the sites I choose will be rather idiosyncratic and incomplete, but they will give you a solid background in statistics and a good idea of areas you might like to explore in more depth.

Don't worry, you won't be left on your own. This is a guided tour. We'll go through it together. I'll explain everything as we go. If I explain too much, move on to the next topic. If I don't explain it enough, reread that section. And if you still don't get it, get online and go to www.statisticssafari.com. You'll discover additional illustrations and practice problems waiting for you there.

This guided tour is primarily conceptual. You don't need to know a lot of mathematics. If you can add, subtract, multiply and divide, you'll be fine. Get yourself a calculator that can do squareroots (that funny symbol $\sqrt{x}$) and you'll really be set. My goal is to help you understand which procedure to use and why. When you know what and why, everything else is quite easy.

I've organized this tour into 10 lessons you can explore at your own pace. Each "day" covers a major area of statistics and is composed of six parts:

Briefly: a quick overview of what's in the chapter.

Introduction: a presentation of new ideas and techniques

Understand: illustrations of key concepts

Remember: facts you need to keep in mind

Do: practice problems and exercises in applying what you know. There are general problems, step by step solutions, and simulations (story problems that require thinking, calculations and interpretation of results.

Summary: a reminder of what you've learned and some items to test your progress. There's a quiz, problems for the current and previous days, and answers.

–Ken Tangen

Additional help and practice problems are available at www.statisticssafari.com

Contents

Introduction   5

Day 1: Measurement   11
Before Collecting Data

Day 2: Central Tendency   29
Describing A Group

Day 3: Dispersion   55
Measuring Diversity

Day 4: Z-Scores   77
Self-Comparisons

Day 5: Correlation   94
Comparing A Group To Itself

Day 6: Regression   115
Predicting The Future And The Past

Day 7: Probability   139
Comparing A Group To A Standard

Day 8: T-Tests   166
Comparing Two Groups

Day 9: 1-Way ANOVA   187
Comparing Three Or More Groups

Day 10: Advanced Designs   211
Testing For Interactions

Practice Final Exam   217
Basic Facts Test   228
Tables   232
Formulas   235
Feedback   237

Introduction

You probably didn't set out to be a number-cruncher. Most people don't. Most are people-persons who spend a great deal of time learning about people, honing their communication skills, and developing intervention strategies. Many don't like math, avoid working with numbers, and never want to do research for a living.

Yet people and numbers seem to go together. The more people there are, the more someone has to keep track of who they are, where they came from, and where they are headed. People, whether they are customers, students or patients, come with test scores, evaluations, and outcome measures. The more you work with people, the more data there is that has to be summarized, understood, interpreted and communicated.

Fortunately, measuring skills don't have to be acquired, just honed. We come equipped at birth with tremendous processing ability. We do it all the time, and with very little effort. When we walk, our brains automatically calculate the slope and curve of the path. We scan for obstructions and estimate the amount of effort needed to climb a hill or jump a curb.

When we listen, we calculate frequency, distance, and quality of tone. We can quickly detect the difference between children who are playing and those who are fighting. When we meet people, our internal measuring systems automatically give us data. Before we even realize it, our predefined questions already have been asked and the data collected. Are they taller than us? Better dressed? Younger, older, or about the same? Would we like being friends; would they like us?

Measuring people is simply a matter of applying the principles we already use to a set of numbers. The only difference is that we must use a more indirect system. Instead of looking directly at a person, we look at numbers which describe that person.

Clearly, our indirect system is not exact. People are not numbers, nor can they be completely described with numbers. There are gaps in our knowledge, holes in our ability to understand. Our measurements are imprecise approximations and subject to errors in data collection, storage, retrieval and interpretation.

To counter the imperfections of our data, we generally restrict our focus to a group of people. Measurement is more like a telescope than a microscope. We do best with a general look. The more detail we try to see, the more errors we make. We're good at broad long-distance generalizations, fairly good at medium levels of specificity, and terrible at close-up inspections.

If we were studying cars, we'd be very good at detecting patterns of traffic, fair at understanding a fleet of cars, and terrible at being auto mechanics. In other words, we would be very good at describing an apple orchard, fair at describing a basket of apples, and poor at understanding a single apple. We can clearly see the mountains on the horizon, we can drive without running into things, but our close-up vision is so poor that we can barely make out the numbers on our cell phone. We can....you get the idea.

Consequently, statistics is the study of groups of people. We look for patterns and trends. We compare one group to another. We interpret our findings as typically true for people in general.

For example, readers of this book generally will not become professional researchers. If you need research done, you'll probably hire someone else to do the job. You're not going to need to recall formulas at a moment's notice but you may need to know where to find them. Even if all you do is watch television, you'll need to know how much faith to put in what researchers tell you.

In other words, studying statistics is a lot like school. You'll be exposed to a lot of material, some of which will stick, most of which will be forgotten. In fact, you wouldn't choose to put yourself through the pain of lectures and homework except for the fact that you'll gain something in the experience that will serve you well the rest of your life.

I've tried to help make the process more efficient. I've prepared some worksheets, quizzes and handouts to help you master the basics of statistics. This collection is not intended as a full-featured textbook but as a supplement to the more traditional presentations. I've given lots of examples but you can stop reading them when it suits you. When you think you understand a topic, take a quiz and check your progress. See if you need more practice problems or if you're ready to move on to the next topic.

I want you to know that I'm only going to be quasi-helpful. I have no intention of taking all of the pain out of activities that will help you. But I do want to do away with the busy work that can bury you. Just as a coach helps you focus your workout—to make sure that your muscle pains are not wasted—think of me as your statistics coach. I'm not going to lower the bar of quality, because I want you to be proud of your success. I am going to provide you with enough help so that you can succeed but the success will be all yours.

What's Stats All About?

Statistics allow us to use numbers to describe what we see. In particular, we typically use statistics to describe a group. Individuals are necessary (they are needed to form a group) but the emphasis is on group data.

Indeed, one of the reasons for using statistics is to reduce a large pile of data down to as few numbers as possible. We are particularly happy when we can reduce group data to a single representative number. This reduction process both summarizes the information and makes it easier for us to communicate it to others. Descriptive statistics are used, then, to make communication simple, clear and easy.

When we describe a group, we organize the data and look for common patterns. A typical pattern in much of the data we collect is that most people have similar enough scores that we can describe the entire group with a single number. Although there is some dispersion (individual differences), it is common to find that most scores bunch in the middle of the variable. This "central tendency" allows us to describe a large group with only one or two numbers.

In addition to description, we can use inferential statistics. We can use statistics to make estimates and predictions from which we infer causation. Inferential statistics tries to answer the basic question: "does the data look like this?"

"Data" of course is any collection of numbers. Whether we ask people's opinion or measure how fast they run, we're going to end up with numbers to crunch. That's data.

"This" in the "does the data look like this?" question is a model. When engineers build bridges or airplanes, they test things out on models: smaller examples of the real thing. In statistics we do the same thing, only our models might be a straight line, a curve, or a another pattern. We are asking if the data looks like our model. If the data looks like our model, we use the model to explain the results. If the data doesn't look like our model, we assume that our results are due to chance.

Chance is a relatively recent philosophical assumption. In ancient times, every detail of life was thought to be caused by something or someone: wizards, poltergeists, or the sun-god, wind-monster or moon-princess. In the story of Jonah and the Great Fish, the guilt of Jonah is determined by casting lots (throwing dice or picking the short straw). The premise was that there was no such thing as a random act.

In more modern times, the emphasis has shifted. The roll of dice is now seen as predictable by the laws of physics. The reason a 6 appears on a die is due to chance, not to the presence of a poltergeist.

In research, we begin with the assumption that whatever pattern we see is due to chance. The reason for this assumption is caution. It is too easy to find spurious relationships and patterns. Our minds are so creative we can see patterns in nearly every ambiguous situation. We see animals in ink blots, a Big Dipper in the stars, familiar faces in clouds, and magical kingdoms in puddles of water.

The desire to avoid such speculation has led to a cautious approach to causation. When we see differences, we begin with the premise that the differences we see are due to chance. We don't infer that the pattern is due to something other than chance until the differences are so large that they can't be ignored. And when we see a pattern, we wait until we're sure the pattern is reliable.

In general, research is limited to 6 comparisons, three for individuals and three for groups. You can compare yourself to yourself, to a group, or to a standard. Self-comparison is simply a matter of counting and charting. Many people track their weight over time with this method. No statistics are usually used; establishing a baseline and tracking performance is often enough.

Comparing yourself to a group often involves grades and test scores. Self-to-group comparisons let you see where you are in relation to everyone else. Entrance test scores (SAT, GRE, LSAT) help describe your general level of achievement.

Comparing yourself to a standard shows your mastery level. Can you jump over a 2-foot high hurdle? How about an 8-foot high wall? Or can you pass the Basic Facts Test included at the end of this book? Self-to-standard comparisons provide hallmarks of performance.

Notice that in the three individual comparisons (self-self, self-group and self-standard) very little statistical knowledge is required. That's because the focus of statistics is on groups, not individuals. Although groups can be compared to a standard, statistics come into play most often when a group is compared to itself or another group.

This introduction to statistics begins by describing a single group. You will likely learn to graph all of the scores and describe its shape. Then you'll learn to identify a score which is representative of the group and try to measure how similar the scores are. Other descriptive tasks will include techniques for comparing an individual to a group. Along the way, you'll learn about probability, causation and logic. You'll discover theories, models, variables and the importance of careful observation. Some time will be spent on levels of measurement and experimental design.

Later, you'll learn about inferential statistics. In general, this aspect of statistics comes down to selecting the proper procedure, doing it step by step, and interpreting the results.

Selecting the Proper Procedure

Clearly, this is the most important thing to learn in statistics. A computer may be able to spit out the results faster than you can type in the data but it is essential to know which procedure to tell the computer to use.

In addition to general information management and summary procedures, there are four specific techniques you'll encounter: correlation, regression, independent t-test and 1-way ANOVA. Each can be used for a different purpose and shouldn't be confused with another.

Correlations and regressions can be used to infer causation. Contrary to popular belief, experiments are not the only procedures used to infer cause-effect. Experiments make us more comfortable in our inferences. When we manipulate one variable and watch another, we're quite comfortable making causal inferences. But when we can't randomly assign people or manipulate events, we can still makes inferences. They are larger leaps of faith but still useful as inference tools. We should be a bit skeptical of experimental findings and a bit more skeptical of correlations.

Correlation and regression are the appropriate tools to use when subjects cannot be randomly assigned to a group. Frequently, the most important social topics must be studied with correlations and regressions. Parents will not allow their children to be randomly assigned to other families for the purposes of studying parenting skills. Instead, data is collected on families as they naturally occur and inferences about parenting skills are made using correlations.

Correlations indicate commonality and are used to measure reliability. Tests of personality, intelligence and achievement are of little value unless their results are reliable (consistent). Typically, a test is given once and then given a week or so later to the same people. In a reliable test, people who scored high the first time should score high the second time around. Similarly, the low scores of the first session should be positively correlated with the low scores of the second testing session.

Regressions are appropriate for making predictions. Predicting future behavior based on past performance is possible if the past behavior is continually improving or declining. The problem with most events, like the stock market, is that past performance is not stable enough to make good future predictions.

Independent t-tests often are used to test the difference between two groups. When subjects are randomly assigned to new and old treatments of the flu, an independent t-test could be used to find out if one group did significantly better (or worse) than the other.

A 1-Way ANOVA is used to compare more than two groups. When subjects have been randomly assigned to three or four groups (different drugs for example), the impact of the drugs on memory could be tested with a 1-Way ANOVA.

But which procedure should you use? By the end of this book, you'll know. You'll be given a hypothetical situation and you will know which procedure is the correct one to use.

Calculating Statistics

If you can follow a recipe, you can calculate statistics. The formulas are straight forward and relatively easy to follow. With practice, anyone can become proficient at calculating statistics with a hand calculator. Although statistics is based on algebraic and mathematical concepts, the actual calculations are simply combinations of adding, subtracting, multiplying and dividing. You need to find the square and squareroot signs on your calculator but advanced functions are not used.

The first few times you calculate a statistic, the process is slow and requires careful attention to detail. But the more you practice the steps, the easier it becomes. At the beginning of a statistics class, students are quiet while they concentrate on doing their calculations. But by the end of the term they can calculate a correlation and chat about what they did over the weekend at the same time. Think of statistics as a language and you'll do fine. As with any language, to be fluent requires practice, practice, and—of course—practice.

A Unique Design

Our "guided tour" is organized to cover the three types of content people can learn: facts, concepts and behaviors.

Facts are the details and minutia that make up much of what we read. Instead of clogging your mind with too many details, I've selected the essential components and summarized them for you. These "Basic Facts" are the details you should remember about statistics. There are no proofs, no lengthy descriptions and no formulas (you can always look them up when you need them). These are the bare facts, an outline of statistics structure.

There also is a conceptual presentation of statistics. It presents the essence of why research is important and how to conduct it. Each concept has multiple illustrations.

And there are things to do. At some point knowledge must be put into practice, and statistics is no exception. Doing calculations is somehow helpful in understanding the process. No one believes you'll want to calculate a correlation by hand but you gain a better understanding of the process by doing problem sets. Even more important is knowing which procedure to select. Our tour gives you practice at choosing the appropriate procedure, applying the formulas to a data set and interpreting the data.

Each "Day"
Briefly
Each day starts with a quick preview of what's in each chapter.

Introduction
The introductions are to give you the gist of the matter. Just as a general understanding of electricity, physics and history help form a cultural literacy, a general understanding of how research is conducted can provide a backdrop for better understanding the world around us. This is a quick and easy explanation of what a statistics textbook and all of its complicated proofs are trying to say.

Understand
Calculating is less important than knowing how to approach a problem and which procedure to use. Computers can do the number crunching but they can't supply the decision making and interpretation. Each lesson covers one basic procedure or process and explains what it is and why you'd want to use it.

Concepts by their nature are easy to keep in mind. Remembering facts and doing behaviors help form a deeper understanding of the concepts. For example, golf is an easy game....conceptually. "To play golf, take a stick and hit a ball into a hole." It's easy to remember, easy to understand, and easy to communicate. Playing golf, however, adds greatly to the understanding of its general principle and of related principles, such as energy, force, angle, and friction.

Similarly, playing statistics—actually doing problems—helps deepen the understanding of what that activity entails. The things you need to remember and the things you need to be able to do are aids to your understanding of the principles of statistics. It is not enough to tell you the rules. It is much better to also give you experiences that reveal a rule's nuances.

Remember

No matter how conceptual the presentation, there is always a fair amount of factual trivia that also must be mastered. I've tried to put all of the information you need to memorize from each lesson in one spot. These "Basic Facts" are things I want you to carry in your head--everything else you can look up if and when you need it.

Do

This book is more than concepts and facts. Each chapter has step-by-step instructions, practice problems to calculate and simulations (story problems of plausible and implausible situations).

Summary

Each chapter ends with a review, a quiz (on that chapter's material) and a progress check (testing your cumulative knowledge). Answers, of course, are provided.

Statistics is an adventure. So put on your walking shoes, grab your bag, and let's

START THE TOUR!

Day 1: Measurement
Before Collecting Data

BRIEFLY

A doctoral candidate (a composite of many graduate students I've met) wanted to hire me as a statistical consultant. I agreed to meet with her and discuss her dissertation project but soon wished I hadn't.

"Here's all my data," she said proudly. She had stacks and stacks of papers piled on the table. "I've finished with data collection and I'm ready for statistics."

"Great. What's your hypothesis?"

"I don't know.. What's a hypothesis?"

"What are you trying to find? What's the purpose of your study?"

"To graduate," she said, confused by my stupidity.

"Aside from being forced to do it, what results did you expect?" I wasn't quite ready to give up.

"That's what you're here to tell me. What procedure should I use?"

"For what? Committing suicide?"

"For the study. I've got all this information, what do I do with it now?"

"Throw yourself on the mercy of your committee and consider applying for law school."

She had no idea what she hoped to learn from the study and yet she had already collected the data. This is equivalent to arriving some place and then deciding whether you should drive, fly or walk there.

She didn't understand that the most important part of research occurs before data is collected. Statistics begins and ends with thinking. In the middle you might do some calculating but the essence of research is thinking.

As with any tour, you have to know where you are planning to go. Even if you're going to explore the entire world, you have to pick a direction. Using a globe or a map to find a beginning region of interest would be a place to start. The same is true of research. You have to know in general both where you are and where you want to be. If you want, you can change course from cancer research to studying the mating habits of groundhogs but you must choose a place to begin your search.

INTRODUCTION

Our first issue concerns what must be done before collecting data. This is pre-number crunching. No math is required! In general, there are five things to ask before collecting data:

1. What Are You Trying To Prove?

Good research begins with good ideas. Fortunately, when we are sitting in our arm chairs and thinking of the way the world functions, we can generate lots of ideas. These constructs (ideas about the way life is) are purely mental abstractions of reality. They can not be directly touched or measured. No one has seen a "self concept" or "personality trait." They are only ideas.

We use our ideas to construct a theory. Theories are composed of constructs. They are collections of ideas; clusters of thought. Theories give us a framework for building our understanding. They inform our inquiries, determine our theoretical questions, and guide our selection of what and whom to study.

There are several criteria for evaluating theories. According to Morgan's cannon (a rule of thumb from the 19th century researcher C. Lloyd Morgan), explanations should be as simple as possible. That is, given an equal choice, our preference is for the simplest explanation. From this general principle of simplicity, we also expect these conceptual collections to have the least amount of contradictions possible. A few contradictions is better than many; no contradictions is best. Put another way, theories should be internally consistent.

Another extension of our love for simplicity is that theories should have a small number of assumptions. Again, there is no absolute number that is acceptable; the smaller the better.

Naturally, we hope these simple explanations are clear and useful. Clarity is better than being vague; but what constitutes clarity is somewhat....vague. Sorry.

Similarly, usefulness is not defined in specific terms but serves as a guideline that reminds us that the hallmark of a theory is not pure truth. The theory that the earth was the center of the universe was useful in its day. It was not the complete truth; nor are all of our current theories completely true. But until we develop better theories, they are useful.

One of the ways theories are useful is that they summarize facts. Although there may be facts which cannot be explained, a theory should summarize as many facts as possible. In that sense, there should be evidence for and against theories. Indeed, theories provide a useful function of contrasting all of the evidence currently available in an area of knowledge.

Usefulness also implies that theories should produce testable hypotheses. It is not enough to have hypotheses; they must be testable. A good theory leads to the creation and testing of many models.

2. What's It Like In Practice

In order to test a theory, we convert it into a model. Theories are pure ideas and abstractions from reality. Like Plato's world of ideas, theories live in their own world and do not necessarily correspond to reality. That's why you hear the expression "It sounded good in theory." Many theories sound good until they are tested for their performance in reality.

Models give us a practical way of testing theories. Models differ from theories in their nature, their scope and their use. By their nature, theories are composed of constructs (ideas). In contrast, models are composed of <u>variables</u>. That is, the basic element of a model is a factor upon which people vary.

After only a few seconds of being with me, you would be able to describe some of my characteristics. Your list might include my sense of humor, my gracefulness (or lack thereof), observations about how I am dressed (shoes tied, Dockers ironed, shirt casual but starched) or my grooming and appearance (hair standing on end, wild look in my eyes, etc.). You might guess my marital status, ethnic background, geographical upbringing and the number of languages I speak.

No matter what is on your list, each item is a variable. It is something on which people vary. Not everyone has my level of musical ability, honesty and silliness. Some people are very musical, some very unmusical; most people, however, are in the middle. Some people are totally dishonest, some totally honest, most are in the middle.

In fact, we believe that in every variable we can measure, there is a middle area in which we will find most people. As a group, people are much alike. Individually, each of us is different in the combination of variables (high on musical ability, low on visual acuity, medium on honesty, etc.). But the variables themselves, when everyone is measured, will show that most people are the same on that variable.

Characteristics that do not vary are called <u>constants</u>. In Einstein's famous $E = mc^2$, c is a constant. But constants are unusual. Particularly in social science, our models are frequently, mostly, almost entirely composed of variables: factors on which people vary.

Models also differ from theories in their scope. Just as model bridges and model trains are smaller, scaled approximations of the real thing, theoretical models often test segments of a theory. Often it is impossible to test a complete model. When the underlying theory is too big, measures are unavailable or intervention is inappropriate, large models are routinely broken into smaller segments and tested separately.

Theories are used to guide research; models are used to test theories. One characteristic of variables is that they are measurable. Because theories are composed of constructs, they are untested theoretical realities. But models are built for the purpose of being tested.

We convert theories to models by operationally defining what we mean. <u>Operational definitions</u> are explanations of what exactly was done. An operational definition for intelligence, for example, could be the score on an intelligence test. Or we could ask people to rate how intelligent they are. Or we might measure brain activity, age or brain weight. Each could be a definition of intelligence. It is up to the experimenter to identify and define what intelligence is in that particular study.

The purpose of operational definitions is clear communication. We want people to know exactly what we do in an experiment so that they could replicate it. We use <u>replication</u> to build confidence in our beliefs. If you and I each do the same experiment in the same way and find the same results, we feel more confident that our findings are valid.

In essence we use models in two ways. We conduct both descriptive and inferential studies. Descriptive studies do not intentionally manipulate the environment; they simply see the world as it is. Our observations are guided by a theory but we do not control what our subjects do or manipulate the environment to ascertain its effect on their behavior.

In contrast, inferential studies have a clear hypothesis and often restrict or control environmental factors. A <u>hypothesis</u> is part of that communication process. We specify what we hope to find, what we expect to find, so we can compare what we see with the model we are testing.

That's an important point: we begin with a model and test it. We do not begin with observations and create a model. We begin with a model of a theory and test it. The reason is that it's easy to decide that we see relationships between events we observe when in fact none exist. This error in judgement is unacceptable; it is like hallucinating: seeing things that don't exist.

To avoid that problem, our theories guide our studies. Theories determine which questions we ask, which variables we include, and how we operationally define factors. Even naturalistic observations are guided by theories. We are very careful. We don't want to infer causation where none exists.

3. Who Is Predicting Whom

Continuous-Discrete Variables

In general, we believe that most variables are <u>continuous</u>. People aren't just smart and stupid, they vary on a continuous scale of intelligence. People are not just rich and poor, their earnings are better described by a continuous variable. Even drug abuse can be considered on a continuous scale (amount of drugs consumed).

Continuous data is a factor which can describe people on a large scale with small steps. But even when the underlying variable is continuous, data can appear <u>discrete</u>. Discrete data is a continuous variable that has been chopped up into parts (high, medium, low; fast, slow).

How a question is worded can change the type of data you collect. "Years of school" is a continuous variable. The answers can range from 0 to 2.3 years to 12 years or more. However, the question "have you ever gone to school?" would result in noncontinuous (discrete) data. Similarly, "Are you employed?" produces discrete information, but the number of days worked is continuous

A discrete variable with only two levels (e.g., yes, no) has its own name: <u>dichotomous</u>.

Independent-Dependent Variables

Traditionally, a distinction is made between independent and dependent variables. It is a characterization based on locus of control. A <u>dependent variable</u> is an outcome. It depends on the performance of the subjects. In contrast, an <u>independent variable</u> is independent of the subjects' control. It is something the researcher selects, manipulates or induces.

The distinction is clear in a traditional experiment: an independent variable is manipulated and a dependent variable is measured. Such a structure provides confidence in making inferences of causation. You stomp on a foot, the person says "ouch." You don't stomp on the foot and the person says nothing. The clear inference is that stomping on a foot causes a person to say "ouch."

Notice that the independent variable is a discrete variable: stomp or not-stomp. It is not measured in continuous increments of pressure but is either there or absent. A variation of this theme is to select high, medium and low levels of an independent variable. But, again, the independent variable is a discrete variable that is manipulated to see what impact it has on a continuous dependent variable.

In many areas of research, variables cannot be directly manipulated, if at all. It would be ridiculous and unethical to assign children to abusive and non-abusive environments to see what impact the independent variable (abuse) has on the dependent variable (self-esteem, for instance). Consequently, the independence of many "independent variables" is in question.

Also, the more complicated models of human behavior include many variables, each impacting and being impacted upon by others. These experimental designs do not lend themselves to the independent-dependent variable distinction. Consequently, there is much to recommend the replacement of independent-dependent variables with the designation of predictor-criterion.

Predictor-Criterion

As an alternative to the independent-dependent variable characterization, the predictor-criterion designation provides more flexibility and more accurately depicts the relationships between model components.

It is more flexible because it includes discrete and continuous variables. Although a discrete predictor (stomp or don't stomp) is good, a continuous predictor would give more information about the amount of pressure needed before you said "ouch."

When it is impossible to manipulate a situation (such as height, gender, or personality type), the term "independent" doesn't aptly describe the variable. Predictors can be discrete (like a traditional independent variable) or continuous (like a correlation or regression). Consequently, a predictor can be an independent or a dependent variable.

The predictor-criterion distinction also is a better description of the relationship between the variables. When subjects cannot be randomly assigned to treatments, the independence of variables is in question. It is clearer to say that a particular variable is being used as a predictor of another.

This distinction also applies to correlational designs. A complex coorelational design might have many variables. Some could be predictors and others criteria. Complex designs might include variables that are both predictors and criteria. These are often called moderator variables (ones that influence only part of a model), intervening variables (variables stuck between a predictor and a criterion) and suppressor variables (variables that filter out noise).

It is important to note that in an actual research study, any variable can be a predictor or a criterion. Annual income, level of education, self-esteem, intelligence—any could be used as a predictor of another. And each could be a criterion. Since the choice is arbitrary, the choice of model components and the hypothesized interrelationships should be determined by the theory being studied.

Although a predictor can be an independent variable or a dependent variable, a criterion is an outcome measure: a dependent variable. Criteria depend on the performance of the subjects. Every criterion is a dependent (measured) variable.

4. Who Are You Going To Study

In addition to deciding what to measure, how to ask the questions and which variables should be predictors and which criteria, the researcher must decide who to study.

Sometimes researchers want to study an entire population: the total number of subjects in a particular area of interest. As the focus of interest changes, the size of the population being studied changes. If you're only interested in what happens to you, the population of interest is one.

Although we think of a population as the number of people in a city or country, in research, a population is any group of interest. It can be the number of people in a family, the number of dogs in a town, or the number of lights on a Christmas tree.

Sometimes the population of interest is too large to measure directly. It is usually not convenient to talk to all of the people in a county or inspect all of the paper clips made daily. When the population is too large, a <u>sample</u> is chosen.

A selected part of a larger group is called a sample. Any group can be thought of as both a collection of smaller groups (a population) and a sample of a larger group. The students in Ms. Mendoza's class are a population to her, a sample of all the 4th graders in the school, a sample of all of the students in the school district, etc.

Obviously, how a sample is chosen determines how well it represents the population. If the first 10 children who enter the class are selected, Ms. Mendoza might have excluded those who rode on the bus (if it ran late that day).

The best way to pick a sample is <u>random sampling</u>. If everyone in the population of interest has an equal opportunity to be selected, the sample is unlikely to be biased in favor of any particular subgroup. This is the ideal. For practical reasons this is seldom done.

Although researchers might want to draw conclusions about the people in general, each person does not have an equal chance of being selected. People living in rural areas—and the disabled, elderly and very young—are generally not included in studies.

What is more practical is <u>convenient sampling</u>. Instead of selecting from a random pool that would include everyone in the world, subjects usually are selected from a convenient pool of people we can convince to be in a study.. In many studies, subjects are selected from those students who are taking introductory psychology, want extra credit and choose to participate in the study.

Whatever size our subject pool turns out to be, researchers must decide how to select them. From our convenient sample, we must choose who to actually run through the maze, give the magic drug, or teach a new technique. A common practice is <u>random selection</u>. With this method, each person in the subject pool has an equal chance of being selected. Being in the pool may not be random but how they are selected from the available subjects is random.

An alternative method is called <u>stratification</u>. When there are certain subgroup comparisons you want to make (male-female, rich-poor or tall-medium-short, for example), subjects are randomly selected from within the categories. First, categories of interest are selected. Then, subjects are randomly selected within each category.

5. What Do The Numbers Mean

Most approaches to research use numbers to measure and describe groups of people. But what meaning do the numbers have?

Obviously, variables do not always use numbers in the same way. You might want to find the average age of a group of people but it's unlikely, for example, that you will want to calculate the average ID number. You know intuitively that averaging ID numbers, room numbers, or Social Security numbers isn't very useful. Such numbers aren't used for their numerical value but simply used as names.

A number which substitutes for a name makes no mathematical assumptions. A marathon runner with a high number on his back doesn't necessarily run faster than one with a small number. The numbers are only used to be able to tell the difference between contestants. Such numbers are at the lowest level of assumption, and are said to be at a <u>nominal</u> level of measurement. It makes no sense to add these numbers together, or find their average; each number is used as a name (nom).

In contrast, the second level of measurement, <u>ordinal</u>, makes two assumptions about its numbers. An ordinal scale distinguishes between members, plus places them in order. Ranking children from tallest to shortest is an ordinal measurement. Winners of a race can be placed in order of 1, 2, and 3 (first, second, and third) but it would be silly to find the average of these numbers. An ordinal scale is like a footrace in a snowstorm: it can tell who came in first but it can't tell how far apart the runners are.

An <u>interval</u> scale includes both of the previous assumptions plus the assumption that the distances between numbers (intervals) are equal. The distance between a score of 8 and a score of 9 on a spelling test is the same distance apart as 3 and 4. Using an interval scale, we could tell the difference between players, find out who came in first, and determine by how much our spelling star won.

Notice, an interval scale assumes equal intervals. In the case of a test, equal intervals means that each item is equally difficult. When the steps are not equal, the scale is ordinal. Consequently, a lot of teacher-made tests look as though they are based on an interval scale but are in fact making ordinal measurements.

The final level of measurement is <u>ratio</u>. A ratio scale includes the previous three assumptions and adds an absolute zero. Because of their absolute zeros, ratio scales have a unique characteristic: they can be used to make ratio comparisons. We can say that a task took twice as long (a ratio of 2 to 1), or that an object weighs a third as much (a ratio of 1 to 3). Our judgments can be described in relation to each other. We can't do that with nominal, ordinal or interval scales.

A zero on a spelling test doesn't mean that the person cannot spell anything at all, only that those selected words couldn't be spelled. The zero is not absolute. Similarly, a zero on a Fahrenheit thermometer doesn't indicate a total lack of heat (if it did we couldn't have minus degrees). In contrast, time, distance, and weight are all ratio scales. A zero on these scales indicates the total absence of that factor.

There are two problems with ratio scales. First, ratio scales are very rare. We often use interval scales (e.g., intelligence scales, reading tests, personality inventories) or ordinal scales (e.g., rating scales), but do not often use ratio scales. We'd like to use ratio scales but we can't find a way to measure personality with a ratio scale.

The second problem is that measurement levels often are ignored. It is common for executives, teachers and others to treat ordinal and interval data as if they were on a ratio scale. Rating scales (1 to 5, 1 to 7, 1 to 10) are ordinal in nature. This is important to understand because some people make the mistake of saying that Group A did twice as well as Group B in the last survey.

When our measurements do not meet the assumptions of a ratio scale, we cannot say that a person with an IQ of 140 is twice as smart as a person with an IQ of 70. Nor can we say that a person who scores 0 on our extroversion scale is not extraverted. These are interval scales.

Interpreting Numbers

The way we measure determines the strength of the conclusions we draw. If we label horses as "jumpers" and "non-jumpers," we have not made any assumptions about which is better, only that they are different. This is a nominal scale. Similarly, if we differentiate between managers and engineers at a nominal level, we make no assumptions concerning which status is best.

At an ordinal level, we could rate horses on their jumping ability or personnel on their sales ability. We could use a scale of 1 to 5, for example. Notice, we could use a two-level scale: thumbs up, thumbs down. The only difference between a two-level ordinal scale and the nominal scale mentioned above is in the assumptions. If we assume that jumpers are better to have than non-jumpers or that sellers are better than non-sellers, the underlying scale is not nominal but ordinal.

Prejudice is a good example of misusing measurement assumptions. Distinguishing between Asians and Whites, north and south, or tall and short is a nominal description. Yet, if underlying our distinctions there is an assumption of one being better than another, we have moved to an ordinal distinction.

It should be clear that the number of spots on an ordinal scale is arbitrary. We are still at an ordinal level when rating on a 10-point scale, a 50-point scale or a 87-point scale. It is the underlying assumptions that determine an item's level of measurement.

To move up to an interval scale in our horse testing, we could set up a course with obstructions for the horses to jump. Again, the number of hurdles included in the course is arbitrary and does not affect the level of its measurement. And, it should also be clear that a score of zero on our hypothetical course does not mean that a given horse cannot jump at all. We may have made all of the jumps too high for any horse to successfully clear.

If we measured how fast each horse ran the course, or how high each one jumped, the measurements would be on a ratio scale. Then, and only then, could we say that one horse jumped twice as high or ran half as fast.

Most of our data is ordinal. When we build a test, we usually don't make each item equally difficult (one of the assumptions for an interval scale). Consequently, our measurements are more like rating scales than precision scientific instruments. Although some of our rating systems are quite complex, the data does not allow us to make fine distinctions between people. We can say one person is more generous, skilled or intelligent than another, but not by how much.

Just as horse-jumping courses usually are composed of items with varying difficulty, items of sales ability differ in difficulty. We do it to save time. With a few items of increasing difficulty, we can distinguish between poor performance and great performance. Without thinking about it, though, we have shifted the underlying level of measurement to an ordinal scale.

This shift is not necessarily bad. It allows us to make gross distinctions with only a few items. But researchers should know which level of measurement they are using. Without such knowledge, they are relying on assumptions which might not be true. We should not fool ourselves into thinking that we are measuring with more precision than is actually present.

Clearly, every level of measurement can be useful. Our tests of increasing difficulty are valuable. We don't have to measure everything on a ratio scale. We can use nominal, ordinal, interval and ratio data. All are useful. Levels of measurement are themselves nominal. One level is not better than another.

UNDERSTAND

Concepts are rules you carry in your head. They are easy to remember and apply to many situations. Here are four measurement concepts that will help you understand the research process. Each illustration describes a variation on the theme.

1. The process flows from the theory

Illustration 1: Linear flow. Think of research as a linear flow from theory to operational definition. This approach starts with a theory. You describe the theory, create a model, select a variable and define that variable in operational terms. It's a process of:

theory-model-variable-definition

By using this approach, your theory protects you from making chance errors. A chance explanation might occur if one day you clapped your hands and the sun came up. It's descriptive but starting with the observation and not a theory allows chance occurrences to have greater weight than they should.

In contrast, if your theory is that you control the sun, the linear approach might lead you to test your theory by clapping one morning and not clapping the next. The tendency with observations alone is that they don't lead to testing your notions; the linear model of research always leads to testing assumptions.

Illustration 2: A circular flow. A different way of looking at the same process is to describe it as a circle. Start anywhere and continue on around the circle. By following the circle, watching the sun come up would lead to revising your theory about controlling the sun, which in turn would produce a new model to test, which would help create a new theory, and so on. The wheel reminds us that science is an on-going process.

Illustration 3: A webbed flow. In practice, research feels interactive. You begin with an idea, think of a procedure, find an interesting fact, argue over coffee. Ideas and methods for testing them seem to come together, in any order, or only after much mental fighting. Research is creative. After years of working with different models, thinking of DNA as a twisted spiral ladder might come as a sudden insight. You might start with a problem and search for its solution. Or you might find a new mathematical procedure and look for ways to use that tool. In practice, the flow of ideas feels both organized and messy. All in all, it feels like this:

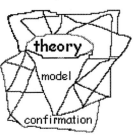

2. All variables are continuous (or can be measured continuously)

Illustration 1: Age is a continuous variable when measured in years, months, days or hours. But when we group people into categories (babies, teens, middle-aged, seniors) we treat it as a discrete variable. Groupings of ages (19-24, 25-39, etc.) also are discrete (everyone in the group receives the same score) but the underlying variable (age) is continuous, not categorical.

Illustration 2: The size of your cereal box can be a discrete variable (large, medium or small) or continuous (net weight).

Illustration 3: When depression is measured by a rating scale (very depressed, slightly depressed, not depressed), the result is a discrete variable. But we could ask the question as a continuous variable: number of statements of sadness the client makes. Similarly, categories of types of depression (dysthymia, major depression, unipolar, bipolar, etc.) would form a discrete variable, but levels of brain chemicals would be a continuous variable.

3. Anything can be a predictor

Illustration 1: Age can predict happiness or happiness can predict age. You can study any potential relationship, many will turn out not to be significant.

Illustration 2: How many beach balls you have could be a predictor of wealth, love of the beach, love of small children or love of odd toys on sale. Unusual variables can be predictors too.

Illustration 3: How fast you drive, how slow you turn, how many cars you own, where you bought your car, and the color, type, make and model of your car, each can be a predictor. The options are endless.

4. Theoretical questions determine which are predictors

Illustration 1: Variables are used to test the model your theory hypothesizes. Without a theory, researchers would study an endless supply of silly variables. The reason no one studies the impact of sunshine on how many wheels a car has is that there is no theory which suggests that it matters. If, however, there was a theory which suggested that sunshine and car design were related, then studying the issue would be logical and appropriate.

Illustration 2: If your theory is that education helps people mature emotionally, your model has two variables: education and maturity. The variables were not picked at random; they were the result of a theoretical relationship.

Illustration 3: Freudian therapists might use displacement and transference as predictors but would not use bilateral stereophonic sound. Behaviorists might use positive reinforcement but would not use identification, and ego strength as predictors. The theory determines which variables are selected.

REMEMBER

Facts are the details of who, what, where and when. Often there are so many facts that we can't remember them all and must look them up. I've tried to collect the facts for each day of the tour in one place. And I've organized them into basic facts (statements you should try to remember), formulas (statements in mathematical form) and terms (vocabulary). Here are the facts about measurement:

Basic Facts:

Theories are composed of constructs.
Models are composed of variables.
Laws have accuracy beyond doubt.
Principles have some predictability.
Beliefs are personal opinions.
Four types of variables: independent, dependent, intervening, and modifying.
Four levels of measurement: nominal, ordinal, interval, and ratio.

Formulas:

For this chapter, there are no complicated formulas for measurement. All that is required is thinking.

Terms:

constant vs variable
construct
convenient pool
continuous vs discrete
criteria vs predictors
dependent vs independent variable
descriptive vs inferential statistics
dichotomous
hypothesis
interval scale
intervening variable
levels of measurement
model vs theory
moderator variable
nominal scale
operational definition
ordinal scale
population vs sample
population of interest
random sampling
random selection
ratio
stratification
suppressor variable
testable hypothesis

DO
Step-by-Step

Research starts with an idea (a construct of a theory), converts that idea into measurable activities (operationally defining the variables), and collects the data. Naturally, the results of model testing impact the theoretical ideas and complete the research cycle. In other words, before you collect any data:

1. <u>Come up with an idea</u>. Ideas come from reading the literature (what other people have done), previous research you've done, and idiosyncratic circumstances (getting hit on the head with an apple).

2. <u>State a hypothesis</u>. The formal statement of a theory's prediction is a hypothesis.

3. <u>Make a model</u>. Convert your idea into something measurable.

4. <u>Pick a population</u>. Decide who to study, how many should be included, and where to find them.

5. <u>Decide on names or numbers</u>. Decide how the numbers you collect are to be interpreted (which level of measurement to use).

My Example

I come up with an idea when reading the newspaper. I see that the NBA basketball finals will be played tonight. I begin thinking about sports, the skill it takes, and how one would learn to become better at hook shots. In the past I've studied pupillary reactivity to emotional stimuli and begin formulating some ideas about how emotions and sports might be linked.

My hypothesis is that anger lowers the accuracy of a player's shot.

My model is that pictures of people playing basketball will reveal strong emotions. I operationally define emotions as the ratings of three judges who are hired to study the photographs of the players.

I choose young children as my population of interest because their faces are so expressive. If a relationship exists, it will show on childrens' faces. An older group might not show their emotions so openly, even if they felt them.

Since judges will be rating the amount of emotion, my results will be measured on an ordinal scale.

YOUR EXAMPLE:

Now it's your turn. Come up with an idea and carry through the design process.

Idea

Hypothesis

Model

Population

Measurement scale:

Practice Problems

1. Operationally define AGE at the nominal, ordinal, interval and ratio levels:

2. Give an illustration of the following:
   a. A theory:

   b. A model:

Simulations

1. In a study which looks at the impact of television on imagination, which is the dependent variable:
   a. television
   b. imagination

2. You work for a pharmaceutical company which is testing a cure for baldness. Subjects are randomly assigned by you into four groups. Each group differs in the number of days they have taken the wonder drug you've created. At the end of each group's treatment, the number of new hairs were counted.

   What is the predictor (independent variable)?

   What is the criterion (dependent variable)?

SUMMARY

Statistics begins with thinking, not calculating. We start with a theory, convert it to a model and decide who, when and how to test it. We decide which variables to measure, how to measure them, and how to collect the data.

Ordinal

Used to report rank or order.

Assumes the numbers can be arranged in order. Allows descriptions of 1st, 2nd and 3rd place but steps need not be the same size. Winning a close race receives the same score as an easy win.

Examples:
  Finish order in contest
  College sports ranking
  Rating scales
  The finish order of your horse.

Nominal

Used as a name.

Makes no mathematical assumption. 0, 12 and 1 have no preference.

Examples:
  The # on a race car
  Bank ID number
  Airplane model #
  Part numbers
  The # on the side of your horse

Ratio

Used to measure physical characteristics.

Assumes 0 is absolute (indicates lack of entity being measured). Allows 2:1, 3:2, "twice as much" and "half as much" comparisons. Zero means no time has elapsed or no distance has been traveled, etc.

Examples:
  Distance, time and weight
  Temperature in Kelvin
  Miles per gallon
  How fast your horse runs

Interval

Used to count conceptual characteristics (IQ, aggression, etc.)

Assumes numbers indicate equal units. Allows distinctions to be made between difficult and easy races but does not allow "twice as much" comparisons. Zero does not mean lack of intelligence, etc.

Examples:
  The # of test items passed.
  Temperature in Fahrenheit
  Temperature in Celsius
  The # of hurdles your horse jumps

1. Which of the following do we select, manipulate, impose or induce:
   a. independent variable
   b. dependent variable
   c. moderator variable
   d. suppressor variable

2. The type of relationships between model components is determined by our:
   a. theoretical questions
   b. empirical analysis
   c. sampling error
   d. statistical bias

3. In an actual research study, annual income could be:
   a. a predictor
   b. a criterion
   c. type I error
   d. either A or B

4. Studies which do not intentionally manipulate the environment are said to be:
   a. experimental
   b. inferential
   c. descriptive
   d. consequential

5. An ordinal scale assumes which of the following:
   a. ordering in magnitude
   b. equal units
   c. A and B
   d. A and B plus other assumptions

6. A test which only orders traits in magnitude is best described as:
   a. cardinal
   b. interval
   c. nominal
   d. ordinal

7. The number of obstacles a horse jumps over would be a(n) _____ measurement.
   a. cardinal
   b. interval
   c. nominal
   d. ratio

8. Selected parts of a larger group are called:
   a. samples
   b. percents
   c. percentiles
   d. population analyses

9. Models are composed of:
   a. variables
   b. percentiles
   c. hodos
   d. confounds

10. The process of converting theories to testable models includes:
   a. confounds
   b. random assignment
   c. random selection
   d. operational definitions

Progress Check

1. As a coach, you are interested in how well your athletes have learned to run. You randomly assigned your athletes into 2 training teams. Team A trained the old fashioned way. Team B used computer-assisted training. After training, you measured the number of minutes needed to run around the track once.

What is the predictor (independent variable)?

What is the criterion (dependent variable)?

2. You work for a vitamin company which is testing a cure for the common cold. Subjects are randomly assigned by you into 5 groups. Each group differs in the number of days they have taken the wonder vitamin you've created. At the end of each group's treatment, the number of coughs were counted.

What is the predictor (independent variable)?

What is the criterion (dependent variable)?

3. Complete the following:

    a. Theories are composed of:

    b. Models are composed of:

    c. Laws:

    d. Principles:

    e. Beliefs:

4. List six criteria for evaluating theories:

    a.

    b.

    c.

    d.

    e.

    f.

5. List 4 levels of measurement:

    a.

    b.

    c..

    d.

Answers

Basic Facts

1. Complete the following:
   a. Theories are composed of: constructs
   b. Models are composed of: variables
   c. Laws: accuracy beyond doubt
   d. Principles: some predictability
   e. Beliefs: personal opinions
2. List six criteria for evaluating theories:
   a. Clear
   b. Useful
   c. Summarizes facts
   d. Small number of assumptions
   e. Internally consistent
   f. Testable hypotheses
3. List 4 levels of measurement:
   a. nominal
   b. ordinal
   c. interval
   d. ratio

Multiple Choice

1. a; 2. a; 3. d; 4. c; 5. a; 6. d; 7. b; 8. a; 9. a; 10. d

Application

11. The predictor was Training and the criterion was Running Time (in minutes).

12. The predictor was Drug Level (number of days treated) and the criterion was number of New Hairs.

Day 2:
Central
Tendency

Describing a group

BRIEFLY

Your tour has landed you on a desert island and two great illustrations of descriptive statistics come to mind: one for social science and one for natural science.

On one hand, you think like a social scientist. Encountering a new species no one has ever seen before--a cross between a Martian and a turtle--you wonder how you are going to describe this historic finding to those back home.

First, you think of Freud (there's something Freudian in that but you let it go). What would Sigmund Freud do in a situation like this? Remembering he was the father of psychoanalysis, you improvise a couch and have the Martles (Turians?) lie down and free associate whatever comes into their minds. Freud's case study approach appeals to you because it doesn't use any numbers.

Second, you recall some learning theorists studied one subject at a time. So you select one Martle and begin a single-subject or $N = 1$ study. You soon realize that having only one individual to study doesn't mean there is a lack of numbers generated. So you want to learn how to organize and plot the data.

The third approach you chose is to study the entire group. There are too many Martles to meet each of them, so you need to find a group representative. Since the goal is description of the entire group, you don't choose the tallest or the shortest. You look for a typical, most common or middle-most representative. So you'll want to know more about the mean, median and mode. You'll want to find the center of a group and why it matters.

On the other hand, the natural scientist in you wants to explore the island itself. It is quickly clear that the island is basically a mountain. Like the pile of stuff that sits on your desk back home: there is one large heap. It has a peak in the middle and symmetrically gets smaller and smaller on each side.

Realizing that I like giving multiple illustrations of the same point, your suspicion is that finding the center of a mountain of data is the same as finding a group representative. Let's see if I accomplish that by the end of this chapter.

The other realization I hope you achieve is that the goal of statistics (as with most of us) is to get the most information with the least amount of numbers.

INTRODUCTION
1. Case Studies

Before trying to change situations, it's often helpful to observe them. This process of naturalistic observation tries to describe the way things are. There is no attempt to change things. Indeed, there is a concerted effort to report the current conditions.

One problem with the process is that it is easy to impact a situation simply by observing it. Often the presence of an observer causes subjects to change their behavior. If someone is watching you walk, do you walk the same way? What happens if people laugh at your jokes or if they read over your shoulder? This observer effect comes from being aware of being watched.

The obvious solution is to make sure that the subjects don't know that they are being observed. But it is not always possible to remain unnoticed. Any group can be naturally observed but ethical and practical considerations limit where, when and how the study can be conducted.

For one subject or a small group, clinicians use case studies. Typically, after each counseling session, therapists write notes on the progress made, goal established and key points to remember. These case notes are often expanded and presented (with identifying information removed) as a case history or case study. Sigmund Freud's theories are based on his clinical observations and case studies. Although Freud was a gifted biological researcher, he is best known for his theoretical and clinical interpretations. Like Freud, most clinicians rely on words, not numbers.

Although case studies tend to personalize observations, they are potentially biased. The validity of the case study relies completely on the interpretations of the clinician making the study. Consequently, findings can vary widely from clinician to clinician. At their best, case studies give enough detail that the reader gets a good feel of the circumstances and the events reported.

Self-reports are case studies which are autobiographical in nature. Typically, the client counts the number of times an event occurs, takes a personal inventory of motives or describes an event in great detail. Self-report checklists, diaries and graphs are widely used by teachers, counselors and parents to track educational and behavioral progress.

2. N = 1 Studies

The experimental version of a case study uses only one subject. These "N = 1 studies" don't use lots of clinical subjects. They study one dog, rat or person at a time.

The German learning theorist Ebbinghaus was the first person to use experiments to study memory. He made long lists of nonsense words and learned them himself. Using himself as his only subject, Ebbinghaus charted the decline of memory over time.

The American learning theorist B.F. Skinner also relied on studying one subject in depth. Skinner avoided complex statistical analyses in favor of counting the number of times a behavior occurred. His a-theoretical approach (no theory) made few assumptions and used replication (rerunning studies) to insure the accuracy of his conclusions. By checking the results of one study as part of conducting a new study, Skinner could show his findings were robust (applies to a lot of situations).

In contrast to case studies, N=1 studies use experimental controls. For example, Skinner created an experimental apparatus to hold his subject and control the amount of light, heat and other environmental factors. The apparatus (colloquially called a "Skinner box" was home for the pigeon, chicken or mouse used in the experiment. Each animal was used repeatedly and each experiment generated a tremendous amount of data about a single subject.

3. Describing A Group
Organize The Data

When you study a group people, one thing becomes quickly apparent: there are a lot of numbers to handle. And the larger the group and the more complex the study, the more data there is.

Like a messy room, data must be collected and sorted until it is easy to handle. Consequently, numbers often are put into rows and columns. This arrangement, formally called a <u>data matrix</u> or data table, is a popular way to organize sales, financial and tax data. Anyone who has used a spreadsheet on a computer is familiar with this row-column organization. Every row is an entity, a person. Every column represents a measurement of some attribute.

For the sake of illustration, let's say that you have eight clients. In order to better understand them, you have measured their age, level of education, their artistic ability and job performance. To protect their identity, ID numbers have been used. A data matrix which profiles this group might look like this:

ID#	Perform	Artistic	Ed.	Age
1	5	1	0	17
2	5	9	12	18
3	4	6	14	23
4	3	5	16	21
5	3	5	15	33
6	3	7	18	42
7	2	4	12	55
8	2	6	14	26

In this example, the first column contains the ID number of each person. The rest of the columns contain a performance rating, the number of items correct on a test of artistic ability, the number of years in school, and each person's age (in years). Each column is a different characteristic, each row a different person.

It is important to note that once data is put into an organized form, it cannot be changed on a whim. Some changes can be made but there are limits. For example, it is permissible to move an entire row or column, because their order is arbitrary. The third column can be moved between the first and second ones without disturbing the structure. Row 7 can be relocated between rows 3 and 4. But the information within a column cannot be changed without its affecting the other information.

If the order of items in one column changes, the other columns must be changed. Since each column describes only one characteristic of an individual, all of an individual's scores must be kept together. In other words, you are free to move an entire person, but you cannot mix-up the component parts.

Regardless of the type of variable, a data matrix lets you scan all of the data at once. People are very good at pattern recognition, better than any computer ever built. We do it quickly and effortlessly. A data matrix makes it easy to see if you remembered to include everything that interests you.

Plot The Data

Some variables can be summarized with a pie chart. Expressed in percentages, each "piece" of the pie represents a segment of that variable. Simple information can be presented quickly with pie charts and they are widely used to summarize information. Researchers use them to show the results of surveys and opinion polls. Businesses use pie charts to show sales by region or by product line.

A bar graph is another categorical graph. The categories are listed from left to right and the values in each category determine the height of each bar. Like pie charts, bar graphs often display percentages.

For more complex information a histogram is can be used. Pie charts show frequencies (how much or how many) but for cumulative non-categorical information a histogram is a better choice. A histogram is a grouped frequency distribution. This imitation of a skyline of a city not only shows frequencies (height on the bars but also shows the relationship between group members. Scores are grouped into sections and the sections are arranged from lowest to highest (left to right).

If this was a graph of age, it would be clear that there are lots of children, lots of elderly, but few in-betweeners. Of course, the price of this clarity is some loss of detail. People who are 62, 63, and 71 are grouped together as if there was no difference between them. Children who are 4, 5, and 8 1/2 also are treated as equals.

Frequency Distributions

Obviously, the shortest description of a group is how many people are in it. So we don't have to write it out each time, this "number" is usually referred to as $\underline{N}$. A group with 10 people would then have a N of 10. N = 123 would describe a group of 123 scores. Unfortunately, N is not very graphic.

A better description of what a group looks like can be found by plotting its scores. This graphic representation of the scores, called a frequency distribution, gives an overview of all of the scores at once. It shows how frequently the same score occurs.

Each score is listed left to right, from lowest to highest. If no one has the same score, it would be represented by a straight line drawn from left to right.

If everyone has the same score, it would be represented by a vertical straight line.

If more than one person has the same score, the line looks more like a mountain or a bell. The number of people who have each score is indicated by the height of the figure. A frequency distribution is also often called a "normal curve."

The Normal Curve

One of the assumptions of science is that variables are normally distributed. That is, there is a pattern to frequency distributions in general. This pattern is consistent no matter what is being measured. Whether you measure mechanical ability, knowledge of world history or tap dancing, when you arrange the scores from low to high they will look like this:

The normal "bell-shaped" curve is a symmetrical distribution. If folded in half the left side looks the same as the right side. Notice that there are more scores in the middle than anywhere else. There are a few people at each extreme, but most are lumped together.

This particular mountain-view has no special significance. It's a matter of convention; a habit developed over the years. A group of people from the side might look like they are standing in line but from the top, you'd see that they are standing in a circle. Arranging people in order of height would look like a staircase from the side and like a straight line from the top. Similarly, our normal curve looks like a mountain from the side. In an aerial view, a normal curve could be drawn more like a pair of lips.

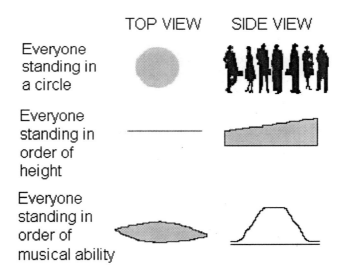

How it is drawn isn't as important as the principle underlying this diagram. Most people are in the middle, with a few on either end. Some of the cars manufactured are extremely reliable, some are lemons. But most are in the middle. Some people are very administratively-, sales-, or musically-talented, some are very untalented, and a whole lot of people are in the middle. Scores on a test of talent, then, could be described as a normal curve.

Skewed Distributions

When a frequency distribution doesn't look like a bell, an alarm should ring. If your people are all the same, you need to look at the data in more detail. There should be differences between individuals. Frequency distributions are reality tests. They remind us that people are generally alike but individuals differ. Most people have the same or similar scores but there will be individual differences. Some people will score low, some high.

Plots, of course, look different each time data is changed. There is some variation. Although two distributions seldom look exactly alike, they usually look like a bell. When the scores don't look like a bell, the distribution is said to be <u>skewed</u>. A skewed distribution has an irregular, asymmetrical shape.

One cause of a skewed distribution is the presence of an outlying score (an unusually high or an unusually low score). An unusually high score impacts the distribution by pulling the high end even farther to the right. This high outlier gives the distribution a long tail on the positive (right) side. The result is a <u>positively skewed</u> distribution.

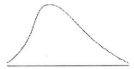

In a <u>negatively skewed</u> distribution, the distribution's tail is pulled toward the negative direction (toward the left side).

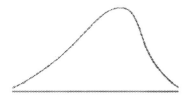

Skewed distributions also occur when a restricted sample is selected from a normal population. Even though the total distribution is normal, taking only a part makes it appear skewed. For example, if we only take the top part of a normal curve, the result would look like a positively skewed distribution.

If we select people who are exceptionally gifted in sales ability, the distribution of their scores would be positively skewed. The hard part is to decide whether the skewed distribution is a false description (due to our pre-selection process) or if it accurately reflects the factor being measured.

Here is another frequency distribution:

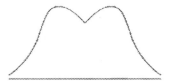

This double-humped camel is a <u>bimodal</u> distribution. Since the height of a distribution is a measure of frequency, this particular irregular shape indicates that a low score and a high score were both more common than the rest of the scores.

Bimodal distributions often are caused by selecting widely-varied groups. This pattern indicates, for example, that you measured young and old clients, and relatively few middle-aged people.

Find A Group Representative

Frequency distributions and graphs give us an overview of the data. But often we want a simple description, a more personalized description. We want to describe the entire group with as few scores as possible.

Sometimes we select a few individuals and highlight their accomplishments. We might choose outstanding teachers, top athletes or best public speakers. Finding an appropriate group representative is a little like a scene in science fiction books or old movies. When the heroes land on Mars they say, "Take us to your leader." Similarly, when confronted with a whole pile of data, we try to find one special score from the entire group.

But there are limits to this approach. The "best" are not good descriptions of what is "typical." When most are followers, finding a leader will not describe the group. Leaders are good group representatives only when most of those in the group are leaders.

Also there are legal and ethical problems in putting someone's picture on the wall and calling them "Mental Patient of the Month." What we need is a numerical description to represent a group of scores. We need to find one score, or make up a hypothetical score, which reflects the nature of the group.

Part of the reason we look for a representative is that we believe that people are more alike than different. On any variable we measure, most people are in the middle of that distribution. If scores were toys, they would not be spread out evenly across a table top. There would not be a line of uniform depth.

4 2 4 6 6 6 10 8 10 8 8 4 2 6 6

They would not be placed in nice neat stacks or columns. There are not two kinds of people: those in Stack A and those in Stack B.

People scores are best described as a heap. There is a pile, a hill or a mountain of scores, and most scores are somewhere in the middle of that pile. There are some scores at each end of the distribution but most scores are in the middle.

That's good news. It means we can describe an entire group of scores with only one or two scores. Because humans are mostly alike on any given variable, we often can describe an entire group of us with a single number.

Most people have about the same amount of musical ability; some are very musical, some are very unmusical, but most are in the middle. The same is true for creativity, aggression, vocabulary and mechanical ability. Consequently, to best summarize a group of scores, we try to find the middle of that distribution, regardless of what it measures.

```
        6
     4  6  8
     2  6  8
  2  4  6  8  10
  2  4  6  8  10
```

There are three ways to find the center of a group: mean, median, and mode. Why three? Because using statistics is like peeking at treasure through tiny, dirty windows. What we see, and how clearly, is determined by which window we use. When the view from all of the windows agree, we can be more certain that what we see is not an illusion. When we see different things from different windows, we are less certain of our observations.

The three major measurements of <u>central tendency</u> are the mean, median and mode. If creatures from Mars came and asked to meet one person most representative of Earth's people, the mean representative would be an average person. The median person would be the one standing in the middle of the group and the mode would be the most popular (most common score).

Mean

The first measure of central tendency, the mean, is more commonly called the average. It is symbolized by a line over the top of variable name; if X is a variable, the mean of X is called X-bar (or bar-X) and is symbolized as X with a bar o $\overline{X}$:

The mean represents the hypothetical, average, typical person. It represents the hypothetical middle point that balances the entire distribution. That's why we end up with 2.4 children or 3.1 cars; they are hypothetical estimates of the middle.

Calculating the mean

To calculate the mean:

    1. Sum (add) all of the scores in a variable

    2. Divide that sum by the number of scores in the distribution.

Calculate the mean of these numbers:

7
6
5
5
5
4
3

The sum of the variable called X is 35. That is:

$$\sum X = 35$$

N (number of scores) is 7.

The mean of these scores is calculated by dividing 35 by 7. So, the mean of these scores is 5. That is:

$$\overline{X} = 5$$

Impact of outlying scores on the mean

Notice that changing an outlying score changes the mean but the median and mode remain the same.

X	Y
7	700
6	6
5	5
5	5
5	5
4	4
3	3

The mean of variable X is 5. But the mean of the Y is 104. Obviously, means are very sensitive to all of the scores in a distribution. Means are very democratic; every score has a say.

Median

In contrast to the mean, the second method of finding the center is not affected by outlying scores. The median ignores the other scores. It doesn't care about their values, only which one is in the middle. Some people think of the median as the team maker, equally dividing the group into two teams with an equal number of scores on each side. Or the barrier running down the middle of a busy street dividing it into equal halves going opposite directions.

When the scores are arranged from low to high, the median is the hypothetical middle-most score. That is, it is the point where 50% of the distribution is above and 50% is below. Since it is a hypothetical point, a median can result in such odd things as 3.5 people or 4.7 buildings, just like a mean.

Calculating a median

Finding the median in a distribution of integers is relatively easy. When there is an odd number of scores: it is the one left over when counting in from either end. When there are an even number of scores, the median is whatever the middle two scores are (if they are the same) or the halfway point between the middle-most two scores when they differ from each other.

Medians are most often used when distributions are skewed. Indeed, when data is presented in medians, ask about the means. If they are quite different, the distribution is highly skewed, and the sample may not be as representative as you would like.

To calculate the median, arrange the scores in order of magnitude from high to low or from low to high (it doesn't matter which one you choose). Select the score in the middle.

In the following numbers, the median is 7:

9
8
7
4
2

What if there's no middle score

The median is the hypothetical middle-most score. If there is no actual middle-most score, the median is the average of the middle two scores of a distribution. So, the median for these numbers is 9:

27
17
15
11
7
5
3
1

Impact of outlying scores on the median

The median is not impacted by outlying scores. It is affected by adding or subtracting a score but not from changing an end score to a larger number. Notice that the median goes unchanged when the surrounding scores are changed.

The median for these scores is 5:

        7
        6
        5
        5
        5
        4
        3

The median of these scores also is 5:

        70
        6
        5
        5
        5
        4
        3

Remember, unlike means, medians are not affected by outlying scores. The median is simply the middle-most point, regardless of who surrounds it

Median: middle of the distribution

The median is the middle of the distribution, not the middle of the raw scores.

The median of the following numbers is 7, not 11:

        14
        12
        4
        11
        3
        7
        5

First the scores must be put in order, then the middle-most score can be found. So, the previous scores would make the following distribution:

        14
        12
        11
        7
        5
        4
        3

The median of this distribution is 7.

Notice that the median of the following data is also 7:

        164
        5
        7
        222
        2

Mode

In addition to means and medians, there is a third way to find a good representative. This third indicator of central tendency is the mode. To people, this score represents the most popular person. To a computer, it is the score most frequently found.

A distribution can have more than one score that appears more often than the rest. In cases where there are two modes, the distribution is said to be bimodal. Anything over two modes is called multi-modal.

Calculating the mode

There are two ways to calculate this popularity. First, the mode may be found by sorting the scores and selecting the one most frequently given. The mode of this distribution is 5:

11
9
5
5
5
2

Second, and more practical in a distribution of many scores, the mode is the highest point on a frequency distribution. If a frequency distribution is accurately drawn, both approaches will yield the same result.

When we draw a distribution, the scores are arranged from left to right, with the lowest scores on the left and the highest scores on the right. When everyone has the same score, the distribution is a straight horizontal line. When more than one person has the same score, the scores are stacked vertically. Consequently, a distribution where everyone had the same score would be represented by a straight vertical line.

Mode: impact of outlying scores

The mode of the following distribution is 5:

5
5
3
4
5
6
7

Note, the mode is not impacted by outlying scores. The mode of the following also is 5:

700
5
5
3
4
5
6
7

Central tendency: why calculate it

Lots of books on statistics tell you how to calculate a mean. This is the only one which explains why. You'll find mathematical presentations, calculator steps and lots of practice problems. But not one of those books explains why anyone would want to calculate a mean. So I'll tell you the secret.

The reason we calculate central tendency is that everything we measure has a center. Every group we select on the basis of chance has a pattern. If you take a bucket of blocks and dump it on the table, they will fall out in a random pattern. The blocks will not stack themselves in neat rows and columns. They will not form a picture or arrange themselves in a circle. Yet there is a pattern.

We believe our universe is orderly. Even chance has a pattern. It's not a pattern of arrangement; it's a pattern of disarrangement. It's a heap, a lump, a mound. A cluster with a center.

We calculate central tendency because there is one. People are organized into those who clap and those who don't. They vary on their ability to clap (jump, dance, sing, surf, type, run....). Yet in their variety there is a tendency for the variable to have a center.

Random variables look like mountains, not stair-stepped plateaus or Stonehenge geometric shapes. People are pretty much alike...regardless of what we measure. There is a commonality to us. There is a center.

We are not lost in the flat desert of data but are faced with a symmetrical mountain. So, it's only natural to figure out how tall the mountain is (mode), how balanced (mean) and where its midpoint (median) is located.

Central tendency: which one to use

As you can see, it matters which number we select to represent a group. Choosing a mean when the curve is skewed gives a false impression. Although sometimes the terms, mean and median, are used as if they were interchangeable, the distinction is an important one.

When the measures of central tendency do not agree, the conclusions drawn from them can be quire different. Our median data may show that we are profitable while the means indicate that we are losing money.

In general, when one is reported, the other should also be given. Having both the means and the medians gives you a better understanding of the data. If they agree, or are close, you can feel confident that either is a good representative of the entire group. When they differ, take a look at the whole distribution.

When a presenter uses means to describe sales, ask about the median values. When medians are used, ask about means. Both measures of central tendency should be close at hand.

UNDERSTAND

Concepts are rules you carry in your head. They are easy to remember and apply to many situations. There are 5 concepts I want to highlight.

1. Get a feeling of what it's like

Illustration 1: Research is a path, but not always a straight one. Sometimes we don't have really clear ideas about what to study or how to ask the right questions. Exploratory research can be helpful too. Find an area of interest and see what's out there. Many researchers will run a little pilot study before devoting all of their time and effort on an area of interest.

Illustration 2: One way to get a feeling for what it's like is to have subjects talk while they participate in your experiment. Verbal protocols are descriptions subjects give about how they are solving a problem. Although not usually objective enough data for making strong predictions, these talk-it-out sessions can show areas where instructions are needed, tasks need to be made easier or other controls need to be added.

2. Not all research uses numbers

Illustration 1: Meta-research, for example, is a special kind of research that uses other people's numbers. Sometimes it is helpful to summarize the state of science in a particular area. These meta-analyses look at all of the studies which have been done in a particular fields (such as learning, personality, etc.) and show how many studies found significant results, how many found no relationship, and where both sides should be applauded and criticized.

Illustration 2: Clinical case studies don't use numbers. Patients are described in terms of their symptoms, unique characteristics and their response to treatment alternatives. The focus is on the symptoms.

Illustration 3: Historical analyses rely on notes, interviews with acquaintances, and the reading of diaries, books, correspondence and historical documents. Psychoanalysis of Abraham Lincoln, for example, would use words, not numbers, to describe his relationship with his mother and the importance of breast feeding and toilet training.

3. The emphasis is on group data

Illustration 1: Although clinicians are interested in how each patient does, in general, research relies on group data. Since we are looking for general principles which apply to all people, there should be consistency in the results. Everyone might not act the same way but general principles assume general patterns of behavior.

Illustration 2: A new drug is thought to be effective if it helps a large group of people, even though it has no impact on your heath.

Illustration 3: Similarly, elections are described by group results, not how an individual votes.

4. One subject is sometimes enough

Illustration 1: Most scientists only study one planet.

Illustration 2: Some anthropologists only study one society or tribe.

Illustration 3: Piaget built his theory of development based on observing one family—his.

5. Central tendency

Illustration 1: A frequency distribution is like a pile of junk. It does not lie flat and organized. The more you pile on, the more it begins to look like a hill.

Illustration 2: A frequency distribution is like a camel: it has a hump...a center.

Illustration 3: A frequency distribution is like an ant hill: it has its highest spot at the center..

Illustration 4: A frequency distribution is like a hut. It is not scattered across the ground but built so that its highest spot is at the center..

Illustration 5: A frequency distribution is not like a flat desert. It is like a mountain on the horizon.

Illustration 6: A frequency distribution is not like a plateau. It is like a sand dune.

REMEMBER

Facts are the details of who, what, where and when. Often there are so many facts we can't remember them all and must look them up. I've tried to collect the facts for each day of the tour in one place. Here are the facts about central tendency.

Basic Facts:

There are 3 measures of central tendency: mean, median and mode.

Formulas:

There are no complicated formulas for central tendency. The mean is what's commonly called an average. It's so simple to calculate, you don't really need a formula. To calculate a mean, simply add up the scores and divide by the number of scores. The symbol for the mean is X (or whatever letter you're using to specify the variable of interest) with a bar over it.

To find the median, find the middle-most score; if there are 2 different scores in the middle, take the average of those two scores. For the mode, find the score that appears most often; find the highest point of a frequency distribution.

Terms:

a theoretical
average
bell-shaped curve
bimodal distribution
case notes; case study
central tendency
data matrix
frequency distribution
histogram
mean
median
mode
N
N = 1 experiment
naturalistic observation
negatively skewed distribution
normal curve
observer effect
outlying score
positively skewed distribution
replication
self-report
skewed distribution
Skinner box
X
SX

DO

Now that we've covered the facts and concepts of central tendency, it's time to put what we know into practice. This section includes Step-by-Step instructions, practice problems, simulations (word problems), quiz and a progress check.

Step-by-Step:

Mean
1. Sum (add) all of the scores in a variable.
2. Divide that sum by the number of scores in the distribution.

Median
1. Arrange the scores in order of magnitude (from high to low or from low to high).
2. Select the score in the middle.
3. If there are two middle scores, add them and divide by 2.

Mode
1. Select the most frequent score

Practice Problems:

Item 1

For the following data, calculate the measures of central tendency:

11
7
9
6
3
10
7
8
5
2
7

N =

$\Sigma X =$

mean =

median =

mode =

Item 2

Calculate the measures of central tendency:

11
7
9
6
2

N =

$\Sigma X =$

mean =

median =

mode =

Item 3

Calculate the measures of central tendency:

10
16
11
6
6
16
3

N =

$\Sigma X =$

mean =

median =

mode =

Item 4

Calculate the measures of central tendency:

24
2
5
11
3
2
2

N =

$\Sigma X =$

mean =

median =

mode =

Item 5

Calculate the measures of central tendency:

800
17
4
2
13

$\Sigma X =$

mean =

median =

mode =

Item 6
Calculate the measures of central tendency:

X
6
6
11
5
6
2

N =

ΣX =

mean =

Item 7
Calculate the measures of central tendency:

2
4
5
7
8
11
5

mean =

median =

mode =

Item 8
Calculate the measures of central tendency:

10
12
8
10
8
10
8
3

N =

mean =

median =

Item 9
Calculate the measures of central tendency:

8
4
2
9
6
4

mean =

median =

Item 10
Calculate the measures of central tendency:

101
13
14
7
7

N =

ΣX =

mean =

median =

Simulations:

Simulation 1

As a realtor, you wonder how much money you typically make per sale ($ in thousands):

6
19
2
6
4
6

Calculate the following:

mean =

median =

mode =

Is the data skewed or normally distributed?

Which measure of central tendency is the best representative for this data?

Simulation 2

As a home buyer, you wonder how much money you'll have to spend fixing up your "new" house. Calculate the appropriate measure(s) of central tendency for these numbers ($ in thousands your neighbors spent on their places):

2
1
2
6
2
26

Which measure of central tendency is the best representative for this data?

Why is this the best representative for this data?

What did you calculate the value to be?

SUMMARY

Our goal is to find a way to summarize a large group of numbers. One part of that process is to find a group's representative. We want one number that will tell us about the entire group. There are 3 basic choices: mean, median and mode. The mean is the hypothetical average person. The median is the middle-most person. The mode is the most popular person. Frequency Distributions can be normal, positively skewed, negatively skewed, constant, flat or bimodal.

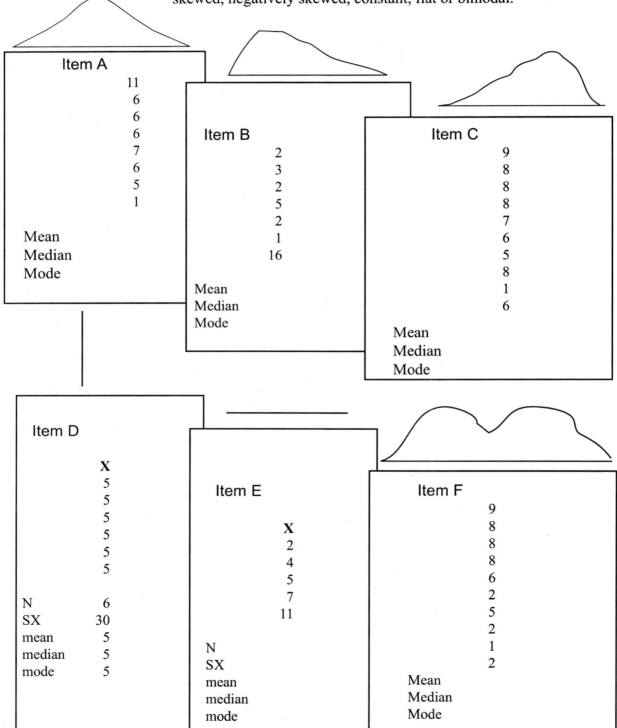

Item A

11
6
6
6
7
6
5
1

Mean
Median
Mode

Item B

2
3
2
5
2
1
16

Mean
Median
Mode

Item C

9
8
8
8
7
6
5
8
1
6

Mean
Median
Mode

Item D

X
5
5
5
5
5
5

N   6
SX   30
mean   5
median   5
mode   5

Item E

X
2
4
5
7
11

N
SX
mean
median
mode

Item F

9
8
8
8
6
2
5
2
1
2

Mean
Median
Mode

1. Which of the following is defined as the middle-most score:
   a. mean
   b. median
   c. mode
   d. frequency distribution

2. Which of the following is defined as the-most common (popular) score:
   a. mean
   b. median
   c. mode
   d. frequency distribution

3. When a curve is skewed, which of the following is the middle-most score:
   a. mean
   b. median
   c. mode
   d. frequency distribution

4. Which of the following is **most** affected by outlying scores:
   a. mean
   b. median
   c. mode
   d. B and C

5. When a distribution is positively skewed, the mean is _____ the median:
   a. lower than
   b. equal to
   c. higher than
   d. twice as large as

For the next five items, consider the following numbers:

76
69
68
67
67
67
66
65
58

6. What is the median of this distribution:
   a. 18
   b. 33
   c. 58
   d. 67

7. What is the mode of this distribution:
   a. 18
   b. 33
   c. 58
   d. 67

8. What is the N of this distribution:
   a. 9
   b. 8
   c. 7
   d. 1

9. This distribution is best described as:
   a. normal
   b. positively skewed
   c. negatively skewed
   d. bimodal

10. Without using a calculator, what is the mean of the above data (hint: see item #9):
   a. 47.5
   b. 64.3
   c. 67
   d. 68.4

Progress Check
1. Complete the following:
    a. Theories are composed of:

    b. Models are composed of:

    c. Laws:

    d. Principles:

    e. Beliefs:

2. List 4 levels of measurement:
    a.

    b.

    c..

    d.

3. Find the mean for these scores:

X
22
24
10
17
3

Mean _____

4. Find the mean, median and mode for these scores:

X
4
4
5
6
4
1

Mean _____

Median _____

Mode _____

5. Find the mean, median and mode for these scores:

$$\underline{X}$$
4
4
9
8
13
4

Mean _____

Median _____

Mode _____

Which is the best measure of central tendency for this group of scores?

6. Find the mean, median and mode for these scores:

$$\underline{X}$$
14
5
5
3
5
2
1

Mean _____

Median _____

Mode _____

Which is the best measure of central tendency for this group of scores?

Answers

Practice Problems

Item 1
N = 11, ΣX = 75, mean = 6.8182 (round it to 6.82), median = 7, mode = 7

Item 2
N = 5, ΣX = 35, mean = 7, median = 7 , mode = no mode (not enough scores; assume it takes 3 of a kind to be a mode)

Item 3
N = 7, ΣX = 68, mean = 9.71, median = 10, mode = none

Item 4
N = 7, ΣX = 49, mean = 7, median = 3, mode = 2

Item 5
ΣX = 836, mean = 167.20, median = 13, mode = none

Item 6
N = 6, ΣX = 36, mean = 6, median = 6, mode = 6

Item 7
Mean = 6, median = 5, mode = none

Item 8
N = 8, mean = 8.63, median = 9, mode = 10

Item 9
Mean = 5.5, median = 5

Item 10
N = 5, mean = 28.40, median = 13

Simulations

Item 1   Mean = 7.17, median = 6, mode = 6. Positively skewed. Median.
Item 2   Median (median = 2; mode = 2 also; mean = 6.50). Negatively skewed.

Multiple Choice:
1. b, 2. c, 3. b, 4. a, 5. c, 6. d, 7. d, 8. a, 9. a, 10. c

Progress Check:
1. Complete the following:
   a. Theories are composed of: constructs
   b. Models are composed of: variables
   c. Laws: accuracy beyond doubt
   d. Principles: some predictability
   e. Beliefs: personal opinions
2. List 4 levels of measurement: nominal, ordinal, interval, ratio
3. Mean = 15.20
4. Mean = 4, median = 4, mode = 4
5. Mean = 7, median = 6, mode = 4. The median is the best measure of central tendency for this group of scores because the distribution is positively skewed. The mean would over estimate the distribution. The mode could also be used but it is less useful than the median
6. Mean = 5, median = , mode = 5. The median is best because.... Ask a friend (or look online).

Day 3: Dispersion

Measuring Diversity

BRIEFLY

Still investigating the lives of Murtles, it is clear to you that not all of them are alike. There is diversity...variability...heterogeneity...dispersion. There is similarity between individuals but they are not all identical. Consequently, in order to describe the entire group, some measure of dissimilarity must also be included. There is a central tendency to the group but there is variety too.

Descriptive statistics is really a group sport. Although some research is done on individuals, most research studies groups. By studying a group, the focus is on what we have in common. A group that has a clear center point can be described by its mean, median and/or mode. Each is a possible representative of the group.

The usefulness of a central representative is not only influenced by if the group has a center but also by how much the scores vary from each other. If everyone has the same score, any of the measures of central tendency can fully represent the group. However, if the scores vary greatly from each other, central tendency is less absolute.

Means, medians and modes do not tell how diverse the scores are. A very homogeneous distribution (very similar scores) and a very heterogeneous distribution (widely varied scores) can have the same mean. But the more varied the scores, the farther they are from the middle, the more difficult it is to summarize a distribution.

Today we tour the following topics:

What is dispersion?

5 ways to measures dispersion

Areas under the curve

INTRODUCTION
What is dispersion?

Dispersion is a measure of how different scores are. It is an inverse measure of cohesiveness. If everyone has the same score, there is no dispersion. The more people who have different scores, the higher the dispersion. If everyone has a different score, dispersion is at its maximum.

Dispersion is calculated by measuring how far every score is away from the mean. If all of the scores are close to the mean, dispersion is low. The more scores differ from the mean, the more dispersion there is. So a group of scores that are tightly clustered around the mean have a low amount of dispersion. A large amount of dispersion indicates the scores are more widely distributed.

When almost no one has the same score, the frequency distribution will be quite wide. There will be more width than height. The mean, median and mode are in the center but there is little agreement among the scores. A distribution with lots of dispersion has little consensus. It is heterogeneous.

When almost everyone has the same score, the frequency distribution will be quite narrow. In a homogeneous group, there is more height than width. In this case, the mean, median and mode are excellent representatives of the entire distribution because almost all of the scores are at or near the center of the distribution.

If the center of a distribution were a statement of truth, dispersion would be a measure of error. When dispersion is high, the errors would be balanced on each side of the argument but there would be little agreement among them. When dispersion is low, the error would be balanced but huddled around the mean.

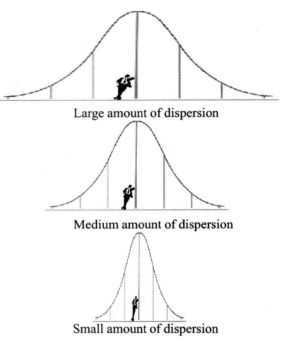

Large amount of dispersion

Medium amount of dispersion

Small amount of dispersion

Five Measures of Dispersion
Range

Like all measures of dispersion, the range of scores gets larger when the distribution of scores is more heterogeneous (dissimilar). The more homogeneous (similar) the scores, the smaller the range.

Range is easy to calculate. It is the highest score minus the lowest score. If the largest score is 12 and the smaller score is 10, range equals 2. If the highest score is 11 and the lowest score is 3, the range equals 8.

Although easy to calculate, range is not terribly helpful for describing a distribution. Without knowing what is being measured, a range of 12 is ambiguous. If we were measuring the number of points each basketball player made during a game, a range of 12 would not be surprising. But if we were measuring the number of goals each hockey player made during a game, a range of 12 would be very unusual.

Range is a good way to check for input errors. If your were inputting scores from a 10-point quiz, a range of 72 would alert you to an input error. The maximum possible in a 10-point quiz is 10 and the lowest possible score is 0, so the range should not be more than 10.

Consider these scores:

15
5
4
4
7
2

The high score is 15 and low score is 2, so the range of these numbers is 13.

Here are some more scores:

121
77
44
155
32
6

The high score is 155 and low score is 6, so the range of these numbers is 149.

Mean Absolute Deviation (MAD)

Like all measures of dispersion, mean absolute deviation (MAD) gets larger when the distribution of scores is more heterogeneous (dissimilar). The more homogeneous (similar) the scores, the smaller the MAD.

Let's break MAD down into its component parts from right to left. The D stands for deviation. MAD is a measure of variation from the mean. To calculate MAD, the mean is subtracted from each score.

In the first column is a variable we'll call X. The mean of this variable is 5. So 5 (column 2) is deducted from each score and the result forms column 3. Since the result is a measure of deviation from the mean, the third column is labeled d (little d).

X	mean	d
7	5	2
5	5	0
5	5	0
5	5	0
3	5	-2

Mean deviation sounds like it should be the mean of those little d's (column 3). We would simply sum the column and divide by the number of scores. But there is a problem. When the little d's are added up, they total zero (2+0+0+0-2 = 0).

But this is to be expected. We started at the mean, which is the balance point of the variable, and measured deviations from it. Since the mean is the center point of the distribution, deviations from it will always add up to 0. So we have two choices. We can take the absolute value of the deviations (which leads us to MAD) or we can square them (as we'll do in Sum of Squares below).

The A of MAD stands for "absolute value" and the M stands for "mean." When we take the absolute value of a number we ignore the sign (positive or negative) of the number. By ignoring the sign, the magnitude of the deviation is added and the result is no longer 0. In the above example, ignoring the positive and negative signs results in a sum of 4 (2+0+0+0+2) and a mean absolute deviation (average of the little d's) of .80 (4 divided by 5).

So MAD (sometime called mean variance) is the average of the absolute values of the deviations from the mean. The mean is subtracted from each raw score and the resulting little d's are averaged (ignoring whether they are positive or negative).

As you can see, MAD is a bit more complicated to calculate than range but more useful as a measure of dispersion. MAD is tied to the mean, gives a quick way to describe dispersion from the mean, and is useful when describing skewed distributions.

The down side is that mean variance doesn't describe the underlying distribution. A MAD of 7 is larger than a MAD of 1.2, but otherwise difficult to interpret.

Sum of Squares

Like all measures of dispersion, Sum of Squares (SS) gets larger when the distribution of scores is more dissimilar (heterogeneous). The more homogeneous (similar) the scores, the smaller the distribution.

Deviation Method

Conceptually, Sum of Squares is an extension of mean variance. Instead of taking the absolute values of the deviations, we square the little critters. It doesn't hurt them. It just gets rid of the negatives. For example:

X	mean	d	d^2
7	5	2	4
5	5	0	0
5	5	0	0
5	5	0	0
3	5	-2	4

Squaring the little d's produces the 4th column. Then just add up (sum) the squared deviations. In this example, the sum of the squared deviations (SS) is 8.

This "deviation method" of calculating Sum of Squares is used to illustrate that it is a measure of dispersion from the mean. After using this method several times, it should be obvious that the deviations are deviating from the mean (since you have to subtract the mean from each score). It also becomes clear that they are squared deviations because you have to square each of the scores. Once this concept is clear, you'll be ready to know the secret: there is an easier way to calculate Sum of Squares.

Raw Score Method

The problem with the deviation method is clear when the mean is not an integer. When the mean is 5, it's not hard to subtract it from every score. When the mean 5.387, it is difficult to know how many places to carry out each of the sub-answers. It's not impossible to do; it's just a pain.

It seems like some mathematician with nothing better to do would have come up with a easier way to calculate Sum of Squares. And, in truth, they did.

The raw score method only uses the raw scores. There are no deviations to calculate. No rounding in the middle of the problem. We just use the X values raw, unprocessed and in the order in which they come. There is no fancy setup.

All we're going to do now is add some numbers (we call it "sum"), square some numbers (multiply a number by itself) and divide. In fact there's nothing more difficult than that in this entire book. Oh, later on we get to push the square root sign but that's more fun than anything else. It usually has its own button on the calculator; it's easy.

I think in visual terms. Here's a map of how I envision calculating Sum of Squares. To me it says add up the first column. I like to call that number Fred. Then square each of the numbers in the second column and add them up. I call this number Ralph. Then square Fred, divide by the number of people in the study and call it Jack. Take Ralph and subtract Jack from it and you have SS (sum of squares).

Now that makes perfect sense to me. I can see me going down the first column adding as I go. I drop the answer on the first dot (Fred). The sideways lines remind me to square and then add, resulting in Ralph (the second dot).

The x and slash beside Fred reminds me to square him and divide by N (number of X scores). And Fred becomes Jack.

Then bring back Ralph, subtract Jack and the result is SS.

As you might guess, not everyone is as gifted at following my maps as I am. My notes always make sense to me but seldom to others. I think of it as the curse of genius.

Some people like math formulas. They think it's easier to communicate how to calculate with Greek letters and funny squiggles. It lacks the grace of my lines and circles but for those traditionalists, here is the formula we use to calculate Sum of Squares:

$$SS = \sum X^2 - \frac{(\sum X)^2}{N}$$

For those who hate math. Relax. It's not as scary as it looks (I could have put a longer scarier version here just to impress you but I'd rather impress you with the simpleness of statistics than with its difficulty). In fact, nothing we are going to calculate is going to be any more difficult than this.

Let's go step by step through the process. It's a four-step recipe.

Assume this is the distribution at issue:

X
11
7
3
4
5
8

First, each number is squared:

X	X²
11	121
7	49
3	9
4	16
5	25
8	64

Second, we sum each column. That gives us 38 for the sum of X and 284 for the sum of X².

Third, square the sum of the first column (38) and divide it by the number of people in the study (N), which is 6. If you're keeping track, 38 times itself = 1444. Then we divide 1444 by 6 (which equals 240.67).

Fourth, subtract what we just calculated (240.67) from the sum of second column (284). That is, 284 - 240.67 equals 43.33.

The Sum of Squares for this distribution is 43.33.

Both Ways At Once

Let's compare the two methods. They will produce the same results but you'll see the raw score method is much easier to calculate. For deviation method, calculate the mean of X. When you add up the X scores, you get 38. And 38 divided by 6 = 6.33.

Subtract the mean from the first X score, put the result in the second column (d). Then square d and put it in the next column (d²). Do it for each X score. And add up the d-squares (d²). The Sum of Squares is 44.33.

X	d	d²
11	4.67	21.78
7	.67	.44
3	-3.33	11.11
4	-2.33	5.44
5	-1.33	1.78
8	1.67	2.78

SS   43.33

For raw score method:
1. Square each X.
2. Sum each column.
3. Square the sum of the first column and divide by N.
4. Subtract the result from the sum of the second column.
Sum of Squares is 43.33

	X	X²
	11	121
	7	49
	3	9
	4	16
	5	25
	8	64
Sum	38	284
N	6	
SS	43.33	

Variance

The fourth measure of dispersion is variance. Like the other measures of dispersion, the larger the variance, the more distributed the scores are. The smaller the variance, the more homogeneous the scores.

Variance of a population is always Sum of Squares divided by N. This is true whether it is a large population or a small one. Regardless of how many scores are in the population, variance is the SS divided by N. Using the numbers from our previous example where the SS was 43.33 and N equaled 6, variance would be 43.33 divided by 6 (which equals 7.22).

Variance of a large sample (N is 30 or more) is also calculated by Sum of Squares divided by N. If there are 40 or 400 in the sample, variance is SS divided by N.

However, if a sample is less than 30, it is easy to underestimate the variance of the population. Consequently, it is common practice to adjust the formula for a small sample. If N is less than 30, variance is SS divided by N-1. Using N-1 instead of N results is a slightly larger estimate of variance and mitigates against the problem of using a small sample. If the above example was not a population but in fact was a small sample, variance would be 43.33 divided by 5, which gives us 8.67.

Obviously, you can't determine whether a group of numbers is a sample or a population just by looking at them. Since a sample is a selected group of scores from a larger group, you must know who you are studying in order to distinguish between a sample and a population. Your family can be a sample of a larger group (all families living in a particular region) or a population (if you don't want to generalize beyond your family). The 50 states can be a population of the United States or a sample of the regional sections of the world.

So to calculate variance, use SS/N for populations (large and small) and for large samples. For small samples, use Sum of Squares divided by N-1. Since most research tends to use large samples of subjects, Sum of Squares divided by N is the most widely used measure of variance.

Standard Deviation

Like the other four measures of dispersion, the standard deviation gets smaller as the scores get more homogeneous and larger the more heterogeneous they become. A small standard deviation indicates the scores are quite similar to the mean; a large standard deviation says the scores vary from the mean.

This measure of dispersion is calculated by taking the square-root of variance. Regardless of whether you used N or N-1 to calculate variance, standard deviation is the square-root of variance. To calculate standard deviation, just push the button on your calculator with this funny symbol on it: $\sqrt{x}$

If variance is 7.22, the standard deviation is 2.69. If the variance is 8.67, the standard deviation equals 2.94.

Technically, the square-root of a population variance is called <u>sigma</u> and the square-root of a sample variance is called the <u>standard deviation</u>. As a general rule, population measures use Greek symbols and sample parameters use English letters. Since we tend to use large samples, we'll focus on the standard deviation.

The standard deviation is "standard" in the sense that it takes steps of equal distance from the mean. Think of it as standing at the mean and taking 3 steps in one direction. It doesn't matter if you step toward the high end or the low end. It only takes three steps to get from the mean to the end of a distribution. If you start at the mean and go toward the positive end, you're there in 3 steps. And it's 3 steps from the mean to the lowest end of the distribution. So the entire distribution is comprised of six steps (3 positive steps and 3 negative steps).

Each of these steps is equal in distance but accounts for a different amount of people. The normal curve is like a mountain. If you're standing on top of the mountain, your first step is always your largest. In a frequency distribution of a normally distributed variable, your first step accounts for the most people. Because most scores are close to the mean, most scores fall within plus or minus one standard deviation from the mean.

In fact, that's our definition of normal. Normal is being close to the mean. Normal musical ability is scoring at the mean plus or minus one standard deviation. Normal basketball throwing is at the mean, plus or minus one standard deviation.

In a normally distributed variable, the percentages in the steps are consistent from distribution to distribution, regardless of what is being measured. Starting from the mean, the first step always accounts for just over 34% of the scores. The next step has 14% and the last step has 2%. Since normal frequency distributions are symmetrical, the percentages work on either side of the mean. So the entire distribution looks like this:

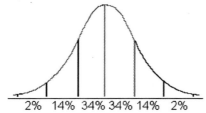

2% 14% 34% 34% 14% 2%

The majority of scores fall at the mean, plus and minus one standard deviation. Normal, then, includes 68% of the scores. So the area from one standard deviation below the mean all the way up to one standard deviation above the mean accounts for 68% of the scores.

Percentages & Percentiles

The nice thing about <u>percents</u> is that they can be from anywhere. You can select the top 1%, the bottom 6%, or the middle 3%. Also, you can randomly select 10% or make sure that the sample is representative of the entire distribution.

Conversely, the problem with percents is that they are not location specific. A sample of 3% can be from anywhere in the distribution.

In contrast, <u>percentiles</u> are location specific. Percentiles are cumulative percentages. Starting from the left of the distribution, percentiles get larger as more and more scores are included. A percentile is the percentage of scores below your score.

If you're at the 50th percentile, 50% of the scores are below you. If you're at the 81st percentile, 81% of the scores are below you. If you're at the 4th percentile, 4% of the scores are below you.

Interestingly, your score doesn't take up any room. If you're at the 50th percentile, 50% are below you and 50% of the scores are above you. A percentile is a hypothetical point and doesn't deduct anything by its presence. Eighty percent of the scores are below the 80th percentile and 20 percent of the scores are above it.

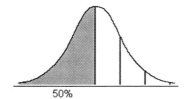 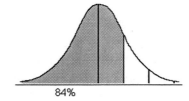

Remember, percentiles are location specific. The 50th percentile is always at the mean, the 80th percentile is always above the mean. Although 2 percent can be from anywhere, the 2nd percentile is always the bottom 2 percent of the distribution. The 98th percentile is everyone but the top two percent. There can be no confusion of a percentile's location.

The problem with percentiles is their interpretation. They sound like equal steps (65, 66, 67, etc.) but the variable being measured is not shaped that way. Percentiles sound like the variable is shaped like this:

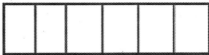

In this example, scores are evenly distributed across the variable. But that's inconsistent with what we know about people. Most people fall in the middle of the distribution, not in equal amounts along its width.

Percentiles give a false impression about the middle of a distribution. Instead of being evenly spaced along a distribution, most percentiles are clustered around the mean. The percentiles in the middle of a distribution are bunched closely together. Although it sounds much better to be at the 60th percentile than the 40th percentile, both positions are virtually at the mean.

Percentiles sound like they are evenly distributed but both the 16th percentile and the 84th percentile are only one standard deviation away from the mean. One standard deviation below the mean is the 16th percentile; one standard deviation above the mean is the 84th percentile. Although easily interpreted (incorrectly) as being evenly distributed, most percentiles are in the middle of a distribution.

Percentiles are not alone in this problem of interpretation; they are merely the most popular. Another approach has been the use of quartiles. Quartiles divide a distribution into 4 parts. Take the median of a distribution. You now have 2 halves. Take the median of each half. You now have 4 quartiles.

Quartiles are composed of a median and two cutoff points. Q1 is the bottom 25 percent; it's at the 25th percentile, Q2 is at the mean, Q3 has the next 25% (and is at the 75th percentile), and Q4 (which you never hear about) is at the far right end of the distribution. Since Q3 is at the 75th percentile and Q1 is at the 25th percentile, 50% of the scores are between Q1 and Q3. This area is called the interquartile range.

Like percentiles, quartiles don't reflect actual differences between scores. And they don't remind us that most people are in the middle of the distribution, with less and less on either side. Quartiles have a lot in common with school grades based on set percentages. Percentiles and quartiles are a lot like grades where an A is 90%, a B is 80% and a C is 70%. They work fine when we want everyone to pass a set standard. A criterion is where everyone is required to spell 100 words correctly, do 20 push-ups or identify 50 state capitals. In these cases, individual performance is judged against a standard, not against other people. With a criterion, all who can spell Mississippi are rewarded equally. There is no comparison between an individual and a group.

Grades, percentiles, and quartiles are not good measures for person-to-group comparisons. They work for comparing individuals to a set standard (able to recite a poem) but don't give a good measure of whether most people can do that skill. Grades, in particular, are difficult to interpret because they are used both as criterion measures and group measures. Sometimes tests are said to be "graded on the curve." That is, the group sets the standard, not the teacher. The "curve" in question is a normal distribution and a C is the middle 50%.

In actual practice, grades don't reflect the middle of the distribution; a C in a course does not necessarily represent the middle of a grade distribution. Many schools worry about "grade inflation," where most students are getting C+'s or B's. In many graduate schools, the average grade is a B+. If the scores are criterion based, it would not be surprising that many students deserved B's. Wouldn't expect that the students at Harvard would be pretty smart? The problem with grades is that it is impossible to tell whether it was a criterion-based or group-based decision.

What we need is a measure that allows us to show where a score is located. What we need is a way to precisely place a score in a distribution. What we need is a way to compare an individual with a group. What we need is a z-score (our topic for tomorrow).

UNDERSTAND

Concepts are rules you carry in your head. They are easy to remember and apply to many situations. There are 5 concepts related to dispersion that I want to highlight.

1. The 2-14-34 Rule

Illustration 1: When you're standing on a mountain, your first step is the largest, so the old saying goes. In statistics, a normal distribution is like a mountain. And if you stand at the peak (where the mean, median and mode are), your first step is the largest. From the mean to the -1 standard deviation accounts for 34% of the scores. Similarly, from the mean to +1 standard deviation holds 34%. The next segments hold 14% each and the furthest away from the mean each hold 2%.

Illustration 2: A frequency distribution is like a triangle. Each side holds 50% in 3 sections: 2% at each end, 14% when getting closer to the mean, and 34% for the section closest to the mean.

Illustration 3: A frequency distribution is like a staircase with 3 steps going up and 3 steps going down. The steps are of unequal size. The first (lowest) steps each hold 2% of the scores. The next steps each holds 14%. The top steps can handle 34% each (or 68% if they are used together).

2. Percents are from anywhere; percentiles start at the bottom

Illustration 1: Percents can be taken from across the distribution. They can be little clusters of blobs or tiny dots systematically selected from the entire distribution. Percents can be the top 10%, the bottom 7% or the middle 12%.

Illustration 2: Five percent is like randomly hitting keys on a piano: they can come from anywhere on the keyboard. The 5th percentile is a specific location: the keys at the far left of the keyboard.

Illustration 3: Percentiles and percents are like the difference between penguins and penguin suits (tuxedos). They sound similar but they mean two different things. Penguin suits (an old name for tuxedoes) can be found anywhere around the world. Penguins are native only to one region. Percentages can be taken from anywhere in a distribution. Percentiles are location specific.

3. Grades, percentiles, and quartiles are not good measures for person-to-group comparisons

Illustration 1: Everyone in the class could get the same score (it was a very smart class).

Illustration 2: Percentiles around the mean are all crowded together because that's where most people are. Out at the ends, percentiles are few and far between.

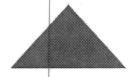

REMEMBER

There are 5 measures of dispersion: range, mean absolute deviation (MAD; mean variance), Sum of Squares, variance and standard deviation. Grades, percentiles, and quartiles are not good measures for person-to-group comparisons.

Formulas:

There are two methods of calculating Sum of Squares: deviation and raw score. The deviation method (which is done to illustrate the concept of dispersion) subtracts the mean for every score, squares the resulting deviations and sums them. When the data includes decimal numbers, the deviation method is very prone to rounding error. The preferred method is the raw score method. It uses this formula:

Terms:

Tangen's 2-14-24 rule
d & d^2
deviation method
dispersion
frequency distribution
interquartile range.
MAD (mean absolute deviation)
N
N-1
percent
percentile
quartile
range
raw score method
standard deviation
Sum of Squares (SS)
variance

DO
Step By Step Instructions
1. Range
From the highest score subtract the lowest score.

2. MAD
First, subtract the mean from each score.

Second, take the absolute value of each deviation. That is, ignore the sign (positive or negative) of each number.

Third, add up the absolute values.

Fourth, divide the sum by the number of scores.

3. Sum of Squares
a. Deviation method
First, subtract the mean from each score.

Second, square each deviation (multiply it by itself).

Third, sum the squared deviations.

b. Raw score method
First, square each raw score.

Second, sum each column.

Third, square the sum of the first column.

Fourth, divide step3 by N

Fourth, subtract step4 from the sum of the second column.

4. Variance
In 3 out of 4 possibilities, variance is SS divided by N.

Use SS/N for large populations

Use SS/N for small populations

Use SS/N for large samples (30 or higher)

If N is less than 30, variance is SS divided by N-1

5. Standard Deviation
Take the square-root of variance.

Practice Problems

Item 1

Calculate the range of the following scores:

19
15
8
5
5
2

High Score _____

Low Score _____

Range _____

Item 2

Assume this is a population and calculate the range and SS of the following scores: Use the raw score method.

11
9
8
7
5
7

ΣX _____

ΣX^2 _____

$(\Sigma X)^2$ _____

N _____

SS _____

variance _____

stdev _____

*For purists, the standard deviation here would be called sigma, because this is a population. I believe the name is less important than the concept, so I use the same words for sample and population.

Item 3

Calculate the variance of the following population:

5
9
7
18
4
2

ΣX _____

ΣX^2 _____

$(\Sigma X)^2$ _____

N _____

SS _____

variance _____

stdev _____

Item 4

Calculate the SS, variance and stdev of the following population:

77
18
9
6
10
60
4

SS _____

variance _____

stdev _____

Item 5

Calculate the range, SS, variance and stdev of the following sample:

12
3
9
5
5
5
7
11
14
- 1

range _____

SS _____

variance _____

stdev _____

Item 6

Calculate the range, SS, variance and stdev of the following sample:

6
2
12
21
7
88

range _____

SS _____

variance _____

stdev _____

Simulations

Simulation 1

As a manufacturer, you want the widgets you produce to all be identical in length. Any variation is considered error. If your criterion is plus and minus 1 standard deviation from the mean, what would be "acceptable" widget length? You are a small manufacturer, here is a the entire population of widgets you've made:

2
18
9
4
51
5
2

$\Sigma X = $ _____

$SS = $ _____

stdev _____

"Acceptable" widget length is from _____ to _____ inches long.

Simulation 2

As a train conductor, you want to know how much variation occurs in the amount of time it takes to collect tickets. Here is a sample of the times:

7
8
4
5
7
3
8

mean _____

median _____

SS _____

variance (remember it's a sample) _____

stdev _____

Simulation 3

You want to know how much variation occurs in amount of time it takes to walk your dog.
Here is sample of the times.

8
6
2
4
4
12

SS _____

variance (remember it's a sample) _____

stdev _____

Simulation 4

As a pilot, you want to know the average number of airline employees "deadheading" with
you on your daily flight to Paris. What limits does it fluctuate between 68% of the time?
Here is a sample of your data:

10
13
12
12
8
6

Calculate the following statistics:

mean _____

median _____

SS _____

variance _____

stdev _____

68% of the time, the number of deadheaders is between _____ and _____.

SUMMARY

To finish off our discussion of measurement, there is a review, quiz, progress check, and chapter answers.

If everyone has the same score, there is no dispersion from the mean. If everyone has a different scores, dispersion is at it's maximum but there is no commonality in the scores. In a normal distribution, there are both repeated scores (height) and dispersion (width).

Percentiles and quartiles imply that distributions look like plateaus. Scores are assumed to be spread out evenly, like lines on a ruler. People are nicely organized in equal-sized containers.

SS, variance and standard deviation imply that distributions look like a mountain. Scores are assumed to be clustered in the middle, people are more alike than different. People are mostly together at the bottom of the bowl with a few sticking to the sides.

You can describe an entire distribution as 3 steps (standard deviations) to the left and 3 steps to the right of the mean. The percentages go 2, 14, 34, 34, 14, and 2. This is believed to be true of all normally distributed variables, regardless of what they measure.

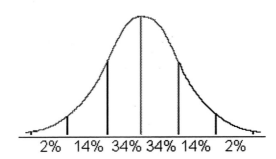

2%   14% 34% 34% 14%   2%

1. The more variability in a distribution, the larger its:
   a. mean
   b. median
   c. mode
   d. standard deviation

2. Sum of Squares is a measure of:
   a. dispersion
   b. central tendency
   c. correlation
   d. interpolation

3. Which of the following is the average of the squared deviations:
   a. mean
   b. mean absolute deviation (MAD)
   c. sum of squares
   d. variance

4. From one standard deviation above to one standard deviation below the mean accounts for what percent of the scores:
   a. 17%
   b. 25%
   c. 34%
   d. 68%

5. Which of the following is the square root of variance:
   a. mean
   b. median
   c. mean variance
   d. standard deviation

6. To calculate the variance of a population, Sum of Squares is divided by:
   a. N
   b. N-1
   c. N-2
   d. N-3

7. Which of the following is the best measure of the fluctuation of a company's stock price:
   a. mean
   b. median
   c. mode
   d. standard deviation

8. When looking for a homogeneous group of people, we should choose the one with:
   a. the largest mean
   b. the smallest mean
   c. the largest standard deviation
   d. the smallest standard deviation

9. A score of 50 on a test with a mean of 100 would be considered normal if the standard deviation was:
   a. 1
   b. 5
   c. 10
   d. 50

10. A score of 50 on a test with a mean of 55 would be considered quite unusual if the standard deviation was:
    a. 1
    b. 5
    c. 10
    d. 20

Progress Check

1. Complete the following:

   a. Theories are composed of:

   b. Models are composed of:

   c. Laws:

   d. Principles:

   e. Beliefs:

2. List 5 measures of dispersion:
   a.

   b.

   c.

   d.

   e.

3. List 2 ways of calculating standard deviations:
   a.

   b.

Answers

Practice Problems

Item 1
High Score = 19
Low score = 2
Range = 17

Item 2
High score = 11
Low score = 5
Range = 6
SX = 47
$SX^2 = 389$
$(SX)^2 = 47^2 = 2209$
N = 6
SS = 20.83
Variance = 3.47
Stdev = 1.86

Item 3
SX = 45
$SX^2 = 499$
$(SX)^2 = 2025$
N = 6
SS = 161.50
population variance = 26.92
standard deviation = 5.19

Item 4
SS = 5249.43
variance = 749.92
stdev = 27.38

Item 5
High score = 14
Low score = -1
Range = 15
SumX = 70
$SumX^2 = 676$
SS = 186
sample variance = 20.67
sample standard deviation = 4.55

Item 6
Range = 86
SS = 5335.33
variance = 1067.07
standard deviation = 32.67

Simulations

Simulation 1
$\Sigma X = 91; SS = 1872$
population stdev (sigma) = 16.35
"Acceptable" widget length is from
- 3.35 to 29.35 inches long

Simulation 2
mean (average) = 6; median = 7
SS = 24. Sample variance = 4
standard deviation = 2

Simulation 3
mean = 6; SS = 64; sample variance = 12.8; stdev = 3.58

Simulation 4
mean = 10.17; median = 11, SS = 36.83, sample variance (N-1) = 7.37, and stdev = 2.71. 68% of the time, the number of deadheaders is between 7.46 and 12.88.

Multiple Choice
1. d, 2. a, 3. d, 4. d, 5. d,
6. a, 7. d, 8. d, 9. d, 10. a

Progress Check
1. Complete the following
   a. Theories are composed of: constructs
   b. Models are composed of: variables
   c. Laws: accuracy beyond doubt
   d. Principles: some predictability
   e. Beliefs: personal opinions
2. Five measures of dispersion:
   a. range
   b. mean absolute deviation (MAD)
   c. Sum of Squares (SS)
   d. variance
   e. standard deviation
3. List 2 ways to calculate standard deviations:
   a. deviation method
   b. raw score method

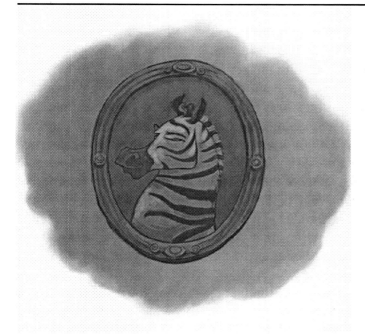

Day 4:
Z-Scores
(Self Comparisons & z-scores)

BRIEFLY

You have learned about operational definitions and levels of measurement.

You've learned 3 measures of central tendency (mean, median and mode). You know that means work great for normal distributions but medians are better for skewed distributions.

You've learned 5 measures of dispersion (range, MAD, Sum of Squares, variance and standard deviation). You know that dispersion is a measure of heterogeneity. A large SS indicates that almost everyone has a different score and a small SS (or range, etc.) shows that the scores are quite homogeneous. You have learned everything there is to know about describing a single group of scores (one variable).

Now we'll explore the individual. Although each single score represents a different individual, by itself one score doesn't reveal much. A score of 95 doesn't indicate anything by itself. It needs context to be properly interpreted. Scoring 95 on a test with 100 items is quite different than 95 out of 12,000. It's the context that allows us to make comparisons.

There are only 3 things you can compare yourself to:

1. You can compare you to yourself. Acting as your own control, you might track your weight, exercise habits, driving skills, piano playing etc. It doesn't matter what others are doing, only how well you are doing compared to your previous experience.

2. You can compare yourself to a standard. You can use pass-fail comparisons for this approach. This is the statistical equivalent of a To-Do list. Just count the number of tasks achieved.

3. You can compare yourself to a group. The mean and standard deviation provide the context to locate a score in relation to the rest of the distribution. Using the mean as a starting point, a score can be expressed in terms of the direction and number of standard deviations it is away from the mean. This process is the statistical equivalent of "Where's Waldo?"

INTRODUCTION

Self to Self

Most self-to-self comparisons don't require any statistics. Usually we simply count and chart our changes. When we want to change our exercise pattern, we might count the number of laps we run or the number of push-ups we do. All in all, it's quite simple.

First, we begin with a baseline. Before making any changes, it's a good idea to see what the current level of performance is. We might keep a diary, a sign-in sheet or a chart, and write in every day how much exercise we did.

Second, we make a change in one aspect of the situation. In our exercise example, we might decide to reward ourselves for every lap around the track. Rewarding ourselves with cookies might not be a good health choice but it doesn't matter to the research design. From a design point of view, all that matters is that we make a change; we change from Condition A to Condition B. And we continue to chart the same behavior. Hopefully, we see an increase in the number of laps we run.

Third, we need to make sure that it's not just time that is making the difference. To do that we switch back from Condition B to Condition A, and continue to chart our progress. If the reward was responsible for the change in behavior, performance will drop back to its original level.

Fourth, once we've eliminated time as an alternative explanation, we return to Condition B.

This ABAB design allows us to test the influence of a situational change on performance. We continuously monitor performance and systematically change the circumstances. No complex statistical analyses are required.

Self to a Standard

Comparing yourself to a standard is the second type self comparison. It also doesn't require much number crunching.

Typically, we set standards by specifying the behaviors which must be visible and the tasks which need to be accomplished. Teachers might require students to memorize a list of words, collect 5 kinds of leaves or identify 32 world rivers. At their best, standards should use 3 principles:

Clear, not vague. Obviously, it is easier to make a comparison to a standard if the criteria are clearly specified. "Clean up your room" is more vague than "Hang up your clothes." Similarly, "You never help out" is more vague than "Take out the garbage."

Increase, not decrease. Usually, it is easier to increase behavior than to decrease it. Consequently, "Eat more carrots" is better than "Eat less junk." Similarly, "Be friendly" is better than "Don't be rude."

Reward, not punish. Giving a reward for good behavior is usually better than punishing bad behavior. Like pulling weeds, punishment stops a behavior but doesn't replace it with anything else. Planting good behaviors is a better plan.

To evaluate whether people meet a standard, a checklist is created. Every accomplished task is checked off the list. Mark all that apply. You've had mumps or not. Performance is evaluated on a yes-no basis The amount of time it takes to mow the backyard is not important, only whether or not the task was completed.

Self to standard comparisons look a lot like To Do lists. You mark off every time you complete a task. Did I wash the car; yes. Feed the cat; yes. Run around the block; no. You use standard in spelling tests, medical profiles, and the Ten Commandments. No complex statistical processes are required. You can simply count how many successes you had.

Self to Group

Comparing yourself to a group is the third type of self comparison. It is a matter of judging your behavior by comparing it to what everyone else does. In contrast to comparing yourself to a standard (did you jump off the roof into the pool; yes or no), self-group comparisons allow "everyone was doing it" explanations. The important thing is whether you did what the group did.

A common self-group comparison is grade equivalents. If you are reading at a 2nd grade level, your performance is equivalent to that group.

Another way of comparing an individual to a group is with percentiles, which are cumulative percentages. You can take the top 5 percent, the middle 23%, the bottom 9% or 2% selected evenly from across the entire distribution. But percentiles are cumulative.

A percentile is the percentage of scores below you. At the 26th percentile, there are 26% of the scores below you. At the 2nd percentile, 2% of the scores are below you and 98% of the scores are above you. As you'll recall, each percentile is a specific point on the distribution; it never varies. In a normal distribution, 50% of the scores are below the mean and 50% of the scores are above the mean. Consequently, the mean is always at the 50th percentile. At the 35th percentile, you are always below the mean, never in the top group. It is a set point.

An even better indicator of score location is called a z-score. Like percentile, z-scores have a set location. A z-score of -1.0 is always in the same spot: one standard deviation below the mean.

Z-scores are the number of standard deviations you are away from the mean. Positive z-scores are above the mean and negative z-scores are below the mean. If z = 0, you're at the mean.

A z-score tells you the number of steps you've taken from the mean and which way you're headed (positive or negative). The standard deviation tells you how big those steps are.

z-score

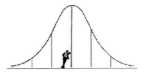

A z-score indicates the distance an individual score is from the mean of a distribution. If a score is at the mean, it has a z-score of 0. Scores above the mean are positive and scores that are located below the mean are negative. In practical terms, z-scores range from -3 to +3. A z of -3 indicates that the raw score is 3 standard deviations below the mean (at the extreme left end of the distribution). A z of 3 indicates that the raw score is at the extreme right end of the distribution.

Since z-scores are expressed in units of standard deviation, they are independent of the variable being measured. A z-score of -1.5 is always one and a half standard deviations below the mean. If z = .5, the score is located at one-half standard deviation above the mean.

Composed of two parts, the z-score has both magnitude and sign. The magnitude can be interpreted as the number of standard deviations the raw score is away from the mean. The sign indicates whether the score is above the mean (+) or below the mean (-). To calculate the z-score, subtract the mean from the raw score and divide that answer by the standard deviation of the distribution. In formal terms, the formula is:

$$z = \frac{X - \overline{X}}{s}$$

Using this formula, we can find z for any raw score, assuming we know the mean and standard deviation of the distribution. What is the z-score for a raw score of 115, a mean of 100 and a standard deviation of 15? First, we find the difference between the score and the mean, which in this case would be 115-100 = 15. The result is divided by the standard deviation (15 divided by 15 = 1). With a z score of 1, we know that the raw score of 115 is one standard deviation above the mean for this distribution being studied.

There are 5 primary applications of z-scores.

Five Z-Score Applications
1. Description

First, z-scores can be used for describing the location of an individual score. What is the z-score for a raw score of 104, a mean of 110 and a standard deviation of 12? 104-110 equals -6; -6 divided by 12 equals -.5. The raw score of 104 is one-half a standard deviation below the mean.

Second, raw scores can be evaluated in relation to some set z-score standard; a cutoff score. For example, all of the scores above a cutoff z-score of 1.65 could be accepted. In this case, z-scores provide a convenient way of describing a frequency distribution regardless of what variable is being measured.

Each z-score's location in a distribution is associated with an area under the curve. A z of 0 is at the 50th percentile and indicates that 50% of the scores are below that point. A z-score of 2 is associated with the 98th percentile. If we wanted to select the top 2% of the individuals taking a musical ability test, we would want those who had a z score of 2 or higher. Z scores allow us to compare an individual to a standard regardless of whether the test had a mean of 60 or 124.

Most statistics textbooks have a table that shows the percentage of scores at any given point of a normal distribution. You can begin with a z-score and find an area or begin with an area and find the corresponding z score. Areas are listed as decimals: .5000 instead of 50%. In order to save space, tables only list positive values. The tables also assume you know that 50% of the scores fall below the mean and 50% above the mean. The table usually has 3 columns: the z score, the area between the mean and z, and the area beyond z.

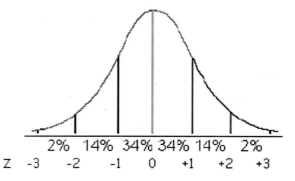

The area between the mean and z is the percentage of scores located between z and the mean. A z of 0 has an area between the mean and z of 0 and the area beyond (the area toward the high end of the distribution) is .5000. Although there are no negatives, notice that a z score of -0 would also have an area beyond (toward the low end of the distribution) of .5000.

Look up z =.1 in Table A at the back of this book. A z score of .1, for example, has an area between the mean and z of .0398. That is, 3.98% of the scores fall within this area. And the third column shows that the area beyond (toward the positive end of the distribution) is .4602. If the z has -.1, the area from the mean down to that point would account for 3.98% of the scores and the area beyond (toward the negative end of the distribution) would be .4602.

Areas under the curve can be combined. For example, to calculate the percentile of a z of .1, the area between the mean and z (.0398) is added to the area below z (which you know to be .5000). So the total percentage of scores below a z of .1 is 53.98 (that is, .0398 plus .5000). A z score of .1 is at the 53.98th percentile.

Third, an entire variable can be converted to z-scores. This process of converting raw scores to z-scores is called standardizing and the resulting distribution of z-scores is a normalized or standardized distribution. A standardized test, then, is one whose scores have been converted from raw scores to z-scores. The resultant distribution always has a mean of 0 and a standard deviation of 1.

Standardizing a distribution gets rid of the rough edges of reality. If you've created a nifty new test of artistic sensitivity, the mean might be 123.73 and the standard deviation might be 23.2391. Interpreting these results and communicating them to others would be easier if the distribution was smooth and conformed exactly to the shape of a normal distribution. Converting each score on your artistic sensitivity test to a z score, converts the raw distribution's bumps and nicks into a smooth normal distribution with a mean of 0 and a standard deviation of 1. Z-scores make life prettier.

Fourth, once converted to a standardized distribution, the variable can be linearly transformed to have any mean and standard deviation desired. By reversing the process, z-scores are converted back to a raw score by multiplying each by the desired standard deviation and adding the desired mean. Most intelligence tests have a mean of 100 and a standard deviation of 15 or 16. But these numbers didn't magically appear. The original data looked as wobbly as your test of artistic sensitivity. The original distribution was converted to z scores and then the entire distribution was shifted.

To change a normal distribution (a distribution of z-scores) to a new distribution, simply multiply by the standard deviation you want and add the mean you want. It's easy to take a normalized distribution and convert it to a distribution with a mean of 100 and a standard deviation of 20. Begin with the z scores and multiply by 20. A z of 0 (at the mean) is still 0, a z of 1 is 20 and a z of -1 is -20. Now add 100 to each, and the mean becomes 100 and the z of 1 is now 120. The z of -1 becomes 80, because 100 plus -20 equals 80. The resulting distribution will have a mean of 100 and a standard deviation of 20.

Fifth, two distributions with different means and standard deviations can be converted to z-scores and compared. Comparing distributions is possible after each distribution is converted into z's. The conversion process allows previously incomparable variables to be compared. If a child comes to your school but her old school used a different math ability test, you can estimate her score on your school's test by converting both to z-scores.

If her score was 65 on a test with a mean of 50 and a standard deviation of 10, her z score was 1.5 on the old test (65-50 divided by 10 equals 1.5). If your school's test has a mean of 80 and a standard deviation of 20, you can estimate her score on your test as being 1.5 standard deviations above the mean; a score of 110 on your test.

UNDERSTAND

Concepts are rules you carry in your head. They are easy to remember and apply to many situations. There are 3 concepts related to z-scores and comparisons that I want to highlight.

1. Compare to Self

Illustration 1: When trying to fix a problem, it's good to get an idea of its current status. Teachers, counselors and parents often start with a baseline of the behavior in question.

Illustration 2: When I weigh myself, I usually say something about how the current number relates to previous measurements. If I've gained 2 pounds or lost 10 pounds, I am always comparing myself to myself.

Illustration 3: I had a friend whose business was losing money. He was sinking into bankruptcy. When things improve slightly, he noted that he was still sinking but slower than he had before.

Illustration 4: Self-comparison is a company saying that they sold more than they did last year.

2. Compare to Standard

Illustration 1: Many people use the rules their parents taught them as standards. They measure personal performance against family history and moral absolutes.

Illustration 2: When doctors use a checklist to diagnose a condition, they are comparing their patients to a standard.

Illustration 3: When a police officer pulls you over for speeding, saying "I drove slower than I usually do" isn't much of defense. Traffic speed is usually compared to a standard: the posted speed limit.

Illustration 4: Comparing to a standard is when a company says that they are selling enough to break even.

3. Compare to Group

Illustration 1: Baserates are for self-comparison, checklists are for standards, and normal curves are for comparing groups.

Illustration 2: For your horse to win, it only has to run faster than the other horses in the race.

Illustration 3: If you enter a talent contest, your singing would be compared to all the other contestants.

Illustration 4: Comparing to a group is when a company says that they are selling better than other companies.

REMEMBER

Facts are the details of who, what, where and when. Here are the facts I've collected for z-scores.

Basic Facts

A z-score indicates how many standard deviations away a score is from the mean.

Z-scores can be used to find an individual, standardize a distribution or set a cutoff.

Formulas

$$z = \frac{X - \overline{X}}{s}$$

Terms

baseline
checklist
cutoff score
grade equivalent
linear transformation
magnitude
normalized distribution
percentile
sign
standardized distribution
standardized score
z-score

DO

Now that we've covered the facts and concepts of z-scores, it's time to put what we know into practice. This section includes Step-by-Step instructions, practice problems, and simulations (word problems).

Step-by-Step

Calculate z-scores

First, find the difference between the raw score and the mean.

Second, divide the result by the standard deviation.

Use z-scores as cutoff scores

A paper company has discovered that short trees make better typing paper. In their plant, they want to set the equipment to automatically reject any trees taller than 1 standard deviation above the mean. Which trees from the following population would be accepted and which rejected?

14
12
11
10
10
10
9
8
6

The mean for this distribution is 10, and the standard deviation is 2.16. Z tells you how many steps to take and in which direction. Since z = 1, you want to add 1 standard deviation to the mean (10 + 2.16 = 12.16). You'd reject any trees as tall as 12.16 feet or taller.

Create a normalized distribution

Frequency distributions based on raw scores often have means and standard deviations that are inconvenient or difficult to handle. To convert a frequency distribution to a more convenient format, raw scores are converted to z-scores and a distribution of z-scores is created. This process is called "standardizing" or "normalizing" a distribution.

Here is a distribution of raw scores:

10
8
7
7
7
6
4

Simply convert each raw score to a z-score. In the raw scores above, the mean is 7 and the standard deviation is 1.69. So 10-7 is divided by 1.69 and the first z score is 1.77.

When all of the raw scores have been converted, there will be a z-score for each raw score:

Raw	z
10	1.77
8	.59
7	0
7	0
7	0
6	- .59
4	- 1.77

This column of z-scores is a distribution of z-scores. The mean will be 0 and the standard deviation will equal 1. And the distribution is now called a "standardized" or "normalized" distribution.

Make a linear transformation

Once a distribution has been "standardized" or "normalized," it's mean and standard deviation can be changed to any desired value. Here are raw scores and their "standardized" distribution.

Raw	z
10	1.77
8	.59
7	0
7	0
7	0
6	- .59
4	- 1.77

First, select the standard deviation you want and multiple each z-score by it. Assuming the new standard deviation was 10, here is how the data would look:

Raw	z	z*stdv
10	1.77	17.7
8	.59	5.9
7	0	0
7	0	0
7	0	0
6	- .59	-5.9
4	- 1.77	-17.7

Second, add the desired new mean to each number. Assuming you wanted the new mean to be 50, here is how the data would look:

Raw	z	z* stdv	+new mean
10	1.77	17.7	67.7
8	.59	5.9	55.9
7	0	0	50.0
7	0	0	50.0
7	0	0	50.0
6	- .59	-5.9	44.1
4	- 1.77	-17.7	32.3

The new distribution now has a mean of 50 and a standard deviation of 10.

Compare two variables

Compare Blake's score on the reading test from her old school to the test used by her new school:

Old Test	New Test
3	22
5	27
7	11
3	31
11	17
13	14
7	21
7	17

Blake's score on the old test was 8; what would her likely score be on the new test? The old test had a mean of 7 and a standard deviation of 3.32. Blake's z-score was .30. The new test has a mean of 20 and a standard deviation of 6.22. Blake's raw score on this test would be 21.87.

Practice Problems

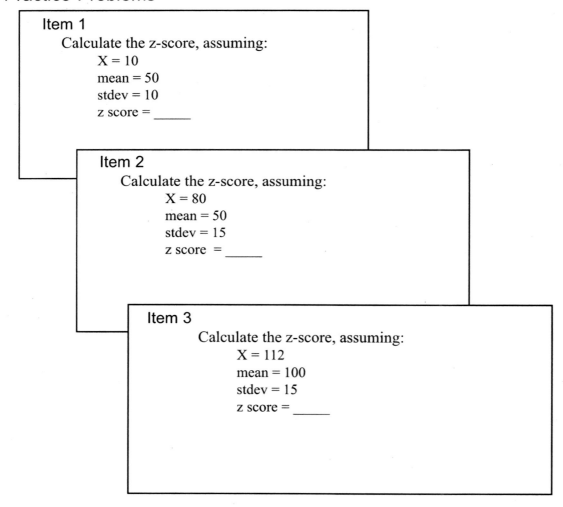

Item 1
Calculate the z-score, assuming:
    X = 10
    mean = 50
    stdev = 10
    z score = _____

Item 2
Calculate the z-score, assuming:
    X = 80
    mean = 50
    stdev = 15
    z score = _____

Item 3
Calculate the z-score, assuming:
    X = 112
    mean = 100
    stdev = 15
    z score = _____

Item 4
Calculate the raw score, assuming:
- $z = 1.5$
- mean = 115
- stdev = 10
- X= _____

Item 5
Calculate the raw score, assuming:
- $z = 2.10$
- mean = 100
- stdev = 20
- X = _____

Item 6
Calculate the raw score, assuming:
- $z = -1.37$
- mean = 100
- stdev = 20
- X = _____

Item 7 (solve for z):
a. What is the z-score for a raw score of 115, a mean of 100 and a standard deviation of 10?

b. What is the z-score for a raw score of 104, a mean of 110 and a standard deviation of 12?

c. What is the z-score for a raw score of 109, a mean of 100 and a standard deviation of 6?

d. What is the z-score for a raw score of 24, a mean of 56 and a standard deviation of 12?

Item 8 (solve for X, instead of z):
a. What is the raw score when $z = 2$, the mean is 100 and the standard deviation is 15?

b. What is the raw score when $z = 1.3$, the mean is 94 and the standard deviation is 12?

c. What is the raw score when $z = -.82$, the mean is 90 and the standard deviation is 11?

Simulations

Simulation 1

As a broker, you are interested in the consistency of XYZ Corp.'s stock price. Your client only wants you to buy this stock when it is unusually low. Consider the following data (assume it is a sample):

6
6
4
6
5
3
3
2
7

What is the sum of Price: _____

What is the median of Price: _____

What is the mean of Price: _____

What is the mode of Price: _____

What is the SS of Price: _____

What is the variance of Price: _____

What is the stdev of Price: _____

What price is at the 84th percentile?

What is the price of the stock when z = 3?

What is the price of the stock when z = -1.3?

What is the price when the stock is at the 95th percentile?

What is the price when the stock is at the 5th percentile?

SUMMARY

A z-score indicates how many steps a person is from the mean. A raw score below the mean corresponds to a negative z score; a score which is above the mean would have a positive z. The standard deviation indicates how big each step is. Approximately 68% of the scores lie within one standard deviation of the mean. That is, a majority of the distribution is from z = -1 to z = +1.

There are 5 primary applications of z-scores:
   a. locating an individual score

   b. using z as a standard. Individual raw scores are converted to z-scores and compared to a set standard. There are two common standards: (a) z = 1.65 and (b) z = 1.96. You'll see these numbers again when we discuss one-tailed and two-tailed tests of significance.

   c. standardizing a distribution and smoothing its data.

   d. making linear transformations of variables; converting the mean and standard deviation to numbers that are easier to remember or handle.

   e. comparing 2 raw score distributions with different means and standard deviations.

1. A z-score of +2 on a test with a mean of 100 and a standard deviation of 10 would equal a score of:
   a. 80
   b. 95
   c. 100
   d. 120

2. Which of the following is calculated by subtracting the mean from the score and dividing by the standard deviation:
   a. intelligence quotient
   b. z-score
   c. t score
   d. b score

3. If a cutoff score is set at z = 1.50, what is the cumulative percentage of scores up to that cutoff:
   a. .063
   b. .437
   c. .737
   d. .933

4. How much area is beyond a z-score of .60:
   a. .027
   b. .229
   c. .274
   d. .374

5. What percentage of scores are between a z-score of -1.3 and a z of +1.65:
   a. 52%
   b. 64%
   c. 68%
   d. 85%

6. A z-score of -1 on a test with a mean of 90 and a standard deviation of 10 would equal a score of:
   a. 80
   b. 95
   c. 100
   d. 102

7. What percentage of scores are between the mean and a z-score of .50:
   a. .0636
   b. .1375
   c. .1915
   d. .8475

8. If z = - 2.60, how much area is "beyond:"
   a. .0047
   b. .1745
   c. .4570
   d. .9045

9. What percentage of scores are between a z-score of -1.05 and a z of +.45:
   a. 53%
   b. 64%
   c. 68%
   d. 72%

10. A z-score of -1.5 on a test with a mean of 100 and a standard deviation of 10 would equal a score of:
    a. 85
    b. 95
    c. 100
    d. 115

Progress Check

1. List six criteria for evaluating theories:

   a.

   b.

   c.

   d.

   e.

   f.

2. List 4 levels of measurement:

   a.

   b.

   c..

   d.

3. Calculate the following using this population data:

$$12$$
$$8$$
$$7$$
$$7$$
$$7$$
$$5$$
$$22$$

Mean _____

Median _____

Mode _____

SS _____

Variance _____

Stdev _____

4. Convert the raw score distribution above to a standardized score distribution.

Answers

Practice Problems

Item 1 $X = 10$, mean $= 50$, stdev $= 10$, $z = -4.00$

Item 2 $X = 80$, mean $= 50$, stdev $= 15$, $z = 2.00$

Item 3 $X = 112$, mean $= 100$, stdev $= 15$, $z = .80$

Item 4 $z = 1.5$, mean $= 115$, stdev $= 10$, $X = 130$

Item 5 $z = 2.10$, mean $= 100$, stdev $= 20$, $X = 142$

Item 6 $z = -1.37$, mean $= 100$, stdev $= 20$, $X = 72.60$

Item 7 (solve for z):

a. $X = 115$, a mean of 100 and a standard deviation of 10?   1.50

b. $X = 104$, a mean of 110 and a standard deviation of 12?   - .50

c. $X = 109$, a mean of 100 and a standard deviation of 6?   1.50

d. $X = 24$, a mean of 56 and a standard deviation of 12?   - 2.67

Item 8 (solve for X, instead of z):

a. When $z = 2$, the mean is 100 and the standard deviation is 15, $X = 130$

b. When $z = 1.3$, the mean is 94 and the standard deviation is 12, $X = 109.60$

c. When $z = -.82$, the mean is 90 and the standard deviation is 11, $X = 80.98$

Simulations

Simulation 1

Price: sum $= 42$; mean $= 4.67$; median $= 5$; mode $= 6$; SS $= 24$

Variance (sample) $= 3.00$; standard deviation (stdev) $= 1.73$

At what price is at the 84th percentile? $[1*1.73] + 4.67 = 6.40$

What is the price of the stock when $z = 3$? $[3*1.73] + 4.67 = 9.86$

What is the price of the stock when $z = -1.3$ $[-1.3*1.73] + 4.67 = 2.42$

What is the price at the 95th percentile? $[1.65*1.73] + 4.67 = 7.52$

What is the price at the 5th percentile? $[-1.65*1.73] + 4.67 = 1.82$

Multiple Choice

1. d, 2. b, 3. d, 4. c, 5. d, 6. a, 7. a, 8. a, 9. a, 10. a

Progress Check:

1. Characteristics of theories: clear; useful; small number of assumptions; summarize facts; internally consistent; testable hypotheses

2. List 4 levels of measurement: nominal, ordinal, interval, ratio

3. Calculate the following population measures:

mean $= 9.71$; median $= 7$; mode $= 7$; SS $= 203.43$

variance $= 29.06$; stdev $= 5.39$

4. Convert the raw score distribution above to a standardized score distribution.

0.42

- .32

- .50

- .50

- .50

- .87

2.28

Day 5: Correlation
Comparing a group to itself

BRIEFLY

When it comes to an individual variable, you have seen all there is to see. You know which level of measurement to use, how to find central tendency (mean, median and mode), how to measure dispersion (range, MAD, Sum of Squares, variance and standard deviation) and how to use a z-score to compare an individual to a group. Having conquered single-variable models, it's now time to explore something more complex.

Correlation is the first 2-variable model we'll consider. Both variables (designated X and Y) are measures obtained from the same subjects. They are pairs of observations on one person. There are twice as many scores but the same number of people. Each column will represent a different variable. Each row will signify one person.

We are still only observing. We are not twisting the tail of an elephant to see what happens. We're watching the group and reporting the findings. We can measure how tall your hut is and how many rocks you own to see if there is a relationship between the two variables. But we're not giving you rocks to see what you do with them.

There are 5 sites to see on this day of our tour:
Correlations
Scatterplots
Causality
Reliability & Validity
Types of correlations
Significance

INTRODUCTION
Scatterplots

To use this simple and yet powerful method of description, we must collect two pieces of information on every person. These are paired observations. They can't be separated. If we are measuring height and weight, it's not fair to use one person's height and another person's weight. The data pairs must remain linked. That means that you can't reorganize one variable (the highest to lowest, for example) without reorganizing the other variable. The pairs must stay together.

Once the data is collected, X is plotted against Y. Make two columns of numbers (X and Y), begin at the top and plot each pair of numbers. Go across the X value and up the Y value. If in the first row of data, X is 5 and Y is 2, go across 5 and up 2 and put a dot. One dot for each pair of numbers results in a graph of scattered dots.

Scatterplots tend to have 1 of 3 patterns. First, the dots might look as if scattered by chance. In this case, dots are everywhere and no particular trend is obvious.

Second, the dots might look like they form a positive trend (from lower left to upper right). This pattern emerges when the low scores on X are paired with low scores on Y.

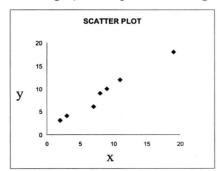

Notice that when X is getting larger, Y usually gets larger too.

Third, the low scores of X might tend to have partners that are high on Y. This trend (from upper left to lower right) indicates a negative pattern. In a negative trend (also called an inverse relationship), when X gets bigger, Y tends to get smaller.

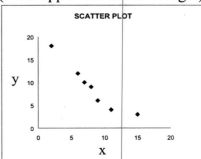

Correlation

Basically, a correlation is a mathematical representation of a scatterplot. Two observations are collected on each subject and the relationship between those two variables is examined. You can compare a group to itself using a correlation. Here are 5 things to note:

Correlations can be used in two ways. First, we can compare the same people on two characteristics. A correlation between eye color and hair color would be such a study. The aim would be to describe the relationship between the two variables for the subjects (people). Two measures of different characteristics might reveal the strength of relationship between the two characteristics.

Second, correlations can be like two snapshots of the same people at different times. A test-retest study would try to describe how reliable a measure is. If you take the temperature of a group of people and re-take it 10 minutes later, you expect a reliable thermometer to provide consistent results. If the test-retest correlation is low, the chances are you have a faulty thermometer.

2 parts

A correlation has both sign and magnitude. The sign (+ or -) tells you the direction of the relationship. If one variable is getting larger (2, 4, 5, 7, 9) and the other variable is headed in the same direction (2, 3, 6, 8, 11), the correlation's sign is positive. In a negative correlation, while the first variable is getting larger (2, 4, 5, 7, 9), the second variable is getting smaller (11, 8, 6, 3, 2).

The magnitude of a correlation is indicated by the size of the number. Correlation coefficients can't be bigger than 1. If someone says they found a correlation of 2.48, they did something wrong in the calculation. Since the sign can be positive or negative, a correlation must be between -1 and +1.

The closer the coefficient is to 1 (either + or -), the stronger the relationship. Weak correlations (such as .13 or -.08) are close to zero. Strong correlations (such as .78 or -.89) are close to 1. Consequently, a coefficient of -.92 is a very strong correlation. And +.25 indicates a fairly weak positive correlation.

Magnitude is how close the coefficient is to 1; sign is whether the relationship is positive (headed in the same direction) or negative (inverse; headed in opposite directions).

Causality

Correlations don't prove causation. A strong correlation is a necessary indicator of causation but it is not sufficient. When a cause-effect relationship exists, there will be a strong correlation between the variables. But a strong correlation does not mean that variable A causes variable B.

In correlations, A can cause B. Or, just as likely, B can cause A. Or, just as likely, something else (call it C) causes both A and B to occur. For a simple example, let's assume that we know nothing about science. But we do notice that when the sun comes up, it gets warm outside. From a statistical point of view, we can't tell which causes which. Perhaps the sun coming up makes it get warm. But it is as likely that when it gets warm the sun comes up. Or the sun and warmth are caused by something else: a dragon (pulling the sun behind it) flies across the sky blowing it's hot breath on the earth (making it warm).

You might laugh at this illustration but think how shocked you'd be if tomorrow it got warm and the sun didn't come up!

It is, of course, perfectly OK to infer causation from correlational data. But we must remember that these inferences are not proofs; they are leaps of faith. Leaping is allowed but we must clearly indicate that it is an assumption, not a fact. Experiments are also leaps of faith but smaller leaps. When conditions are controlled, we feel more confident about leaping to conclusions but the responsibility for inferences is ours. Research doesn't prove, it indicates, supports and tends to suggest inferences. Surveys don't prove people feel something; they indicate that people report feeling that way.

Reliability & validity

Although correlations can't prove cause and effect, they are very useful for measuring reliability and validity. Reliability means that you get the same result every time you use a test. If you're measuring the temperature of liquid and get a reading of 97-degrees, you would expect a reliable thermometer to yield the same result a few seconds later. If your thermometer gives different readings of the same source over a short period of time, it is unreliable and you would throw it away.

We expect many things in our lives to be reliable. When you flip on a light switch, you expect the light to come on. When you get on an elevator and push the "down" button, you don't expect the elevator to go sideways. If you twice measure the length of a table, a reliable tape measure will yield the same result. Even if your measuring skill is poor, you expect the results to be close (not 36 inches and then 4 inches). You expect the same results every time; you expect reliability.

Reliability, then, is the correlation between two observations of the same event. Test reliability is determined by giving the test once and then giving the same test to the same people 2 weeks later. With this test-retest method, you would expect a high positive correlation between the first time the test was given and the second time.

A test with a test-retest reliability of .90 (which many intelligence tests have) is highly reliable. A correlation of .45 shows a moderate amount of reliability, and a coefficient close to zero indicates the test is unreliable. Obviously, a negative test-retest reliability coefficient would indicate something is wrong. People who got high scores the first time should be getting high scores the second time, if the test is reliable.

There are 3 basic types of reliability correlations. A test-retest coefficient is obtained by giving and re-giving the test. A "split half" correlation is found by correlating the total score for the first half with the total score for the second half for each subject. A parallel forms correlation shows the reliability of two tests with similar items.

Correlations also can be used to measure validity. Although a reliable test is good, it is possible to be reliably (consistently) wrong. Validity is the correlation between a test and an external criterion. If you create a test of musical ability, you expect that musicians will score high on the test and those judged by experts to be unmusical to score low on the test. The correlation between the test score and the expert's rating is a measure of validity.

Validity is whether a test measures what it says it measures; reliability is whether a test is consistent. Clearly, reliability is necessary but not sufficient for a test to be valid.

Calculating Correlation

The formula used to calculate correlation looks scary but isn't. It's got all kinds of letters and symbols. It looks very impressive. It's sort of the King Kong of formulas. But like an old horror movie, once you've seen it over and over again, and understand its parts, there's nothing scary about it. Old movies with simple special effects become comical because we can see the wires and recognize the painted backdrops. So don't let the formula for the Pearson r (the one we use to calculate correlation) freak you out. We'll go step by step through the whole process until its tame and familiar.

Types of Correlation

Two Types of Relationship

Correlations can test both linear and monotonic relationships. In a monotonic relationship, the 2 variables move in the same direction but at different rates. Ordinal data is monotonic; both variables move from 1st place to 2nd place but in different sized steps. If one variable goes from 6 to 8 and the other variable moves from 6 to 17, the relationship is monotonic but not linear.

In a linear relationship, two variables move at the same rate. That is, a linear relationship forms a straight line, while a monotonic zigzags in one direction.

Three Kinds of Correlations

In general, correlations can be classified by the type of data they use. In a Pearson r, two continuous variables are used. For example, a correlation between age (in years) and height (in inches) would involve two continuous variables.

In contrast, a phi (pronounced "figh", as in fee, phi, foe, fum) correlation uses two discrete variables. If age and height were measured as discrete variables (high, medium and low), the correlation would be a phi correlation.

A point-biserial correlation uses one continuous and one discrete variable. A correlation between age (in years) and whether you voted in the last election (yes-no) would be a point-biserial correlation.

Significance

It is possible to test a correlation coefficient for significance. A significant correlation means the relationship is not likely to be due to chance. It doesn't mean that X causes Y. It doesn't mean that Y causes X; or that another variable causes both X and Y. Although a correlation cannot prove which causes what, the correlation coefficient (r) can be tested to see if it is likely to be due to chance.

First, determine the degrees of freedom for the study. The degrees of freedom (df) for a correlation are N-2. If there are 7 people (pairs of scores), the df = 5. If there are 14 people, df = 12.

Second, enter the statistical table "*Critical Values of the Pearson r*" (see Table B at the end of this book) with the appropriate df. Let's assume there were 10 people in the study (10 pairs of scores). That would mean the degrees of freedom for this study equals 8.

Go down the df column to eight, and you'll see that in order to be significant with this few of people, the magnitude of the coefficient has to be .632 or larger. Notice that the table ignores the sign of the correlation. A negative correlation of -.632 or larger (closer to -1) would also be significant.

Evaluate r-squared

A correlation can't prove that A causes B; it could be that B causes A...or that C causes both A & B. The coefficient of determination is an indication of the amount of relationship between the two variables. It gives the percentage of variance that is accounted for by the relationship between the two variables.

To calculate the coefficient of determination, simply take the Pearson r and square it. So, .89 squared = .79. In this example, 79% of the variance can be explained by the relationship between the two variables. Using a Venn diagram, it is possible to see the relationship between the two variables. It is the area of overlap.

To calculate the amount of variance that is NOT explained by the relationship (called the coefficient of non-determination), subtract r-squared from 1. In our example, $1-r^2 = .21$. That is, 21% of the variance is unaccounted for.

UNDERSTAND

Comparing a group to itself

Illustration 1: Using a correlation in a test-retest study is like using a mirror twice. Look once, look twice and compare the results. You can compare one image of the group to another image of the same group.

Illustration 2: Using a correlation in a test-retest study is like a family reunion. You see how much they look like they did last year.

Correlation

Illustration 1: A correlation is like a toothache: it's a hint. Your toothache might be caused by a cavity. The symptom is a hint but not proof. Similarly, correlations hint at relationships but don't reveal cause and effect.

Illustration 2: A correlation is like gossip. If you see a couple having dinner together you don't know what the truth is. It might indicate that he invited her, that she invited him, or that they both are guests of someone else.

Significance

Illustration: Significance is like clarity. Without clarity, you can't be sure what you see. But when the window is clean and clear, you are sure you are seeing something. You still can't tell whether what you see is cause or consequence or chance. The view might be of truth or fiction but the window is clean.

Coefficient of determination

Illustration: Measuring correlations is like measuring house size. Instead of square footage, correlations have squared area of relationship. The larger the coefficient of determination (r^2), the more relationship between the two dependent variables. If $r^2 = 1.00$, 100% of the variation of one variable can be explained by the variation of the other variable. When 100% of the house's square footage is accounted for, there is nothing else to find. We completely understand how much influence one variable has on another.

Coefficient of non-determination

Illustration: Correlations also include measuring the part not explained by the relationship between variables. The coefficient of non-determination (what we don't understand) is our percentage of stupidity. When $1-r^2 = 1.00$, we are 100% stupid. Fortunately, the higher the coefficient of determination, the smaller the coefficient of non-determination. When r^2 is high, $1-r^2$ is low.

REMEMBER

Facts are the details of who, what, where and when. Here are the facts I've collected for correlations.

Basic Facts

There are 3 major types of correlations: Pearson (2 continuous variables), phi (2 discrete variables) and point biserial (1 continuous and 1 discrete variable).

Formulas:

$$r = \frac{\sum XY - \frac{(\sum X)(\sum Y)}{N}}{\sqrt{(\sum X^2 - \frac{(\sum X)^2}{N})(\sum Y^2 - \frac{(\sum Y)^2}{N})}} \quad \text{or more simply it is seen as: } r = \frac{SSxy}{\sqrt{SSxSSy}}$$

Terms:

coefficient of determination
coefficient of nondetermination
correlation
correlation coefficient
df
linear
magnitude
monotonic
negative correlation
parallel forms
Pearson r
phi
point-biserial
positive correlation
r
r^2
reliability
scatterplot
sign
significance
split half
test-retest
validity

DO
Step-by-Step
1 Calculate SSx
To calculate the SSx, we use this formula:

$$SS = \sum X^2 - \frac{(\sum X)^2}{N}$$

2. Calculate SSy
Use the same formula, substituting Y for X.

3. Make a new variable: XY
We create a new variable by multiplying every X by its Y partner. So this:

X	Y
11	10
7	9
3	2
4	6
5	5
8	8

becomes this:

X	Y	XY
11	10	110
7	9	63
3	2	6
4	6	24
5	5	25
8	8	64

4. Calculate SSxy
This is an unusual Sum of Squares. All the other SSs are measuring dispersion from a mean, so they can't be smaller than zero (everyone is at the mean). But since we've created this variable, the SSxy can be either positive or negative.

First, sum the XY's. In our example, the Sum of XY = 292.

Second, multiply the Sum of X by the Sum of Y. That is, 38 times 40 = 1520.

Third, divide the result of Step 2 by N (the number of scores). So, 1520 divided by 6 = 253.33

Fourth, subtract the result of Step 3 from the result of Step 1. And 292 minus 253.33 = 38.67. So the SSxy = 38.67.

5. Find r (correlation coefficient)

Here is the formula we use:

$$r = \frac{SSxy}{\sqrt{SSxSSy}}$$

First, multiply the SSx times SSy. In our example, that is 43.33 times 43.33, which equals 1877.49.

Second, take the square-root of Step 1. The result in this case, of course, is 43.33.

Third, divide the SSxy by the result of Step 2. So, 38.67 divided by 43.33 =.89.

The Pearson r = .89

6. Test significance of r

First, determine the degrees of freedom for the study. The degrees of freedom (df) for a correlation are N-2. Since there are 6 people in the current example, df = 4.

Third, enter the table at df = 4. Go down the column to 4 and across to the .05 alpha level (2 tailed). In our table (Critical Values of the Pearson r), only the .05 alpha values are listed (we're trying to make this easy!).

Fourth, find the critical value the table lists. In this example, the critical value = .811.

Fifth, compare the r you calculated with the book. If your r is larger than the book, you win: the correlation is significant and the relationship is not likely to be due to chance. Since the coefficient you calculated (.89) is larger than the critical value (.8111), the correlation in our example is significant.

7. Evaluate r-squared

The coefficient of determination is an indication of the amount of relationship between the two variables. It gives the percentage of variance that is accounted for by the relationship between the two variables.

To calculate the coefficient of determination, simply take the Pearson r and square it. So, .89 squared = .79. In this example, 79% of the variance can be explained by the relationship between two variables.

To calculate the amount of variance that is NOT explained by the relationship (called the coefficient of non-determination), subtract r-squared from 1. In our example, $1-r^2$ = .21. That is, 21% of the variance is unaccounted for.

An example

We create a new variable by multiplying every X by its Y partner. So this:

X	Y
2	17
13	3
10	4
3	18
2	19
12	11

becomes this:

X	Y	XY
2	17	34
13	3	39
10	4	40
3	18	54
2	19	38
12	11	132

Then, we sum each column. The sum of X = 42, the sum of Y = 72, and the sum of XY is 337.

Calculate the SS for X (136) and the SS of Y (256).

And calculate the SS of XY. Multiply the sum of X by the sum of Y (42 * 72 = 3024). Now divide the result by N (the number of pairs of scores = 6); 3024/6 = 504. Subtract the result from the Sum of XYs (337-504 = -167).

Notice the SSxy is negative. It's OK. The SSxy can be negative. It is the only Sum of Squares that can be negative. The SSx or the SSy are measures of dispersion from the variable's mean. But we created the XY variable; it's not a real variable when it comes to dispersion. The sign of SSxy indicates the direction of the relationship between X and Y.

So we have a negative SSxy because X and Y have an inverse relationship. Look at the original data: when X is small (2), Y is large (17). When X is large (13), Y is small (3). It is a consistent but inverse relationship. It's like pushing the yoke down and the plane going up.

Let's finish off the calculation of the Pearson r. Multiply the SSx by the SSy (136 * 256 = 34816). Take the square root of that number (sqrt if 34816 = 186.59). Divide the SSxy (-167/186.59 = -.895). Rounding to 2 decimal places, the Pearson r for this data set equals -.90. It is a strong, negative correlation.

Practice Problems

Item 1

What is the correlation between these two variables

X	Y
11	11
5	3
7	8
6	7
2	3
14	13

$SSx =$ _____

$SSy =$ _____

$SSxy =$ _____

$r =$ _____

$r^2 =$ _____

Item 2

What is the correlation between these two variables:

Laughter	Statistics
11	19
3	6
5	6
15	12
7	8
6	7
9	14

SSx _____

SSy _____

$SSxy$ _____

$r =$ _____

Item 3

How much variance is accounted for between these two variables:

Strength	Speed
10	3
9	7
6	9
3	11
1	15
7	6

SSx = _____

SSy = _____

SSxy = _____

r = _____

r^2 = _____

Item 4

What is the correlation between these two variables:

Strength	Peace
2	2
6	5
7	8
5	19
9	12
9	8
4	5

r = _____

r^2 = _____

Item 5

How much variance is **NOT** accounted for between these two variables:

Strength	Anxiety
2	3
9	7
3	4
1	1
7	12

$r =$ _____

$r^2 =$ _____

$1-r^2 =$ _____

Item 6

Is there a significant relationship between the number of petals on a flower and how much the elephant plant weighs:

Petals	Weight
18	1
15	7
8	11
5	6
4	19
2	22

$r =$ _____

$r^2 =$ _____

Item 7

What percent of variance do love and peace have in common:

Love	Peace
12	1
22	3
8	0
18	1
2	11
9	4
2	5
7	23

$r =$ _____

$r^2 =$ _____

Simulations

Simulation 1

As a political pollster, you wonder if there is a relationship between what people paid for their car and their annual salary (in $10,000). You measure everyone in the country (it's a very small country) and here is the resulting data:

Salary	Car Cost
1	2
2	5
14	17
11	8
7	6
3	4

What is the sum of Salary: _____

What is the SS of Salary: _____

The variance of Salary is: _____

What is the mean of Car: _____

What is the SS of Car Cost: _____

Since you are interested in commonality, which of the following tests should you perform:
    a.t-test
     b.ANOVA
    c.correlation
    d.regression
    e.multiple regression

Perform the comparison you selected in the item above. Select only the appropriate one(s). What was the result of your calculation?

    a =

    b =

    r =

    t =

    F =

Calculate the coefficient of determination for this data.

Simulation 2

As a researcher, you are interested in the relationship between depression and sugar. You have measured several patients at the local hospital on each variable, and now hope to find how related these two variables are.

Depression	Blood Sugar
7	3
8	3
11	2
6	8
7	7
3	12
1	14

What is the sum of Depression:_____

What is the SS of Depression: _____

The stdev of Depression is: _____

What is the median of Sugar: _____

What is the SS of Sugar: _____

Since you are interested in commonality, which of the following tests should you perform:
  a.t-test
  b.ANOVA
  c.correlation
  d.regression
  e.multiple regression

Perform the comparison you selected in the item above. Select only the appropriate one(s). What was the result of your calculation?

  a =

  b =

  r =

  t =

  F =

How much joint variance is accounted for?

Simulation 3

As a governor, you are interested in how similar are the patterns of serving in the state legislature and being rich. You have measured each representative on each, and now hope to find how related these two variables are. The numbers below represent the number of years in the legislature and net worth (in billions) of each member.

Years	Net Worth
5	2
10	14
8	5
10	16
5	13
3	8
1	5

What is the SS of Years: _____

What is the variance of Years: _____

What is the SS of Net Worth: _____

What is the stdev of Net Worth:_____

What is the SSxy: _____

Since you are interested in commonality, which of the following tests should you perform:

    a.t-test
    b.ANOVA
    c.correlation
    d.regression
    e.multiple regression

Perform the comparison you selected in the item above. Select only the appropriate one(s). What was the result of your calculation?

    a =

    b =

    r =

    t =

    F =

What are the degrees of freedom (df) for this study?

What is the critical value for this statistic?

Is there a significant relationship between Years & Net Worth at the .05 alpha level?

SUMMARY

To measure the strength of relationship between two variables, it would be best to use a correlation.

A correlation can only be between -1 and +1.

The closer the correlation coefficient is to 1 (either + or -), the stronger the relationship.

The sign indicates the direction of the relationship.

The coefficient of determination is calculated by squaring r. The coefficient of determination shows how much area the two variables share; the percentage of variance explained (accounted for).

The coefficient of nondetermination is calculated by subtracting the coefficient of determination from 1. The coefficient of nondetermination shows how much the two variables don't share; the percentage of unexplained variance.

To calculate the correlation between two continuous variables, the Pearson product-moment coefficient is used. To calculate the correlation between two discrete variables, the phi coefficient is used. To calculate the correlation between one discrete and one continuous variable, the point biserial coefficient is used.

Correlations are primarily a measure of consistency, reliability, and repeatability.

Correlations are based on two paired-observations of the same subjects.

A cause-effect relationship has a strong correlation but a strong correlation doesn't guarantee a cause-effect relationship. In a correlation, A can cause B or B can cause A or both A and B can be caused by another variable. Inferences of cause-effect based on correlations are dangerous. A correlation shows that a relationship is not likely to be due to chance but it cannot indicate which variable was cause and which effect.

Test-retest coefficients are correlations.

In order to make good predictions between two variables, a strong correlation is necessary.

1. Which of the following gives the correlation between two discrete variables:
   a. phi
   b. Pearson r
   c. least squares criterion
   d. point biserial

2. Which of the following correlation coefficients shows the greatest amount of relationship:
   a. .23
   b. .45
   c. .71
   d. -.89

3. A correlation between two variables:
   a. proves A causes B
   b. proves B causes A
   c. proves C causes A
   d. none of the above

4. Which of the following best describes this correlation:

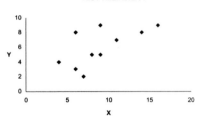

TEST RELIABILITY

   a. strong, positive
   b. strong, negative
   c. weak, positive
   d. weak, negative

5. Which indicates the percentage of explained variance:
   a. normalized coefficient
   b. standardized coefficient
   c. coefficient of determination
   d. coefficient of nondetermination

6. Which is used to calculate the correlation between two continuous variables:
   a. phi
   b. Pearson r
   c. point-biserial
   d. 1-Way ANOVA

7. Correlations are used as measures of:
   a. reliability
   b. discreteness
   c. continuation
   d. dispersion

8. Correlations assume that:
   a. subjects are randomly assigned
   b. confounds are obvious
   c. 2 independent variables are equal
   d. none of the above

9. Correlations have both sign and:
   a. power
   b. magnitude
   c. independence
   d. egocentricity

10. Correlations are numerical expressions of:
    a. confounds
    b. thoughts
    c. dispersion
    d. scatterplots

Progress Check

1. List five measures of dispersion:
   a.

   b.

   c.

   d.

   e.

2. List three types of correlation and the kind of variables with which they are used:

   a.

   b.

   c.

3. Calculate the following characteristics of people's closets:

Hats	Gloves
15	13
10	5
7	5
7	5
2	2
4	3

The sum of Hats: _____

SS of Hats: _____

Mean of Gloves: _____

Population variance of Gloves:_____

SSxy: _____

r =

df =

The critical value for r =

Is there a significant relationship between Hats & Gloves at the .05 alpha level?

Answers

Practice Problems

Item 1
SSx = 93.50
SSy = 83.50
SSxy = 84.50
r = .956
r^2 = .915

Item 2
SSx = 98
SSy = 145.43
SSxy = 85
r = .712
r^2 = .507

Item 3
SSx = 60
SSy = 87.50
SSxy = - 69
r = - .95
r^2 = .907 (round to .91)

Item 4
SSx = 40
SSy = 189.71
SSxy = 31
r = .356 (round to .36)
r^2 = .127 (round to .13)

Item 5
SSx = 47.20
SSy = 73.20
SSxy = 47.20
r = .80
r^2 = .64
$1-r^2$ = .36

Item 6
SSx = 207.33
SSy = 326
SSxy = - 211
r = - .81
r^2 = .66
$1-r^2$ = .34

Item 7
SSx = 354
SSy = 414
SSxy = - 155
r = - .40
r^2 = .16
$1-r^2$ = .84

Simulations

Simulation 1
sum of X = 38
SSx = 139.33
variance of x = 23.22
mean of Y = 7
SSy = 140
correlation
r = .90
r^2 = .81 (coefficient of determination)

Simulation 2
sum of X = 43
SSx = 64.86
stdev of X = 3.29
median of Y = 7
SSy = 132
SSxy = - 87
correlation
r = - .94
r^2 = .88 (coefficient of determination; shared variance accounted for)

Simulation 3
SSx = 72
population variance of X = 10.29
SSy = 172
population stdev of Y = 4.96
SSxy = 66
correlation
r = .59
df = 7-2 = 5
critical value = .755
not significant at .05 alpha level

Multiple Choice
1. a, 2. d, 3. d, 4. c, 5. c,
6. b, 7. a, 8. d, 9. b, 10. d

Progress Check

1. List five measures of dispersion:
   a. Range
   b. Mean Absolute Variance
   c. Sum of Squares
   d. Variance
   e. Standard Deviation

2. List three types of correlation and the kind of variables with which they are used::
   a. Pearson r 2 continuous variables
   b. Phi 2 discrete variables
   c. Point biserial 1 continuous and 1 discrete variable

3. Calculate the following characteristics of people's closets:
   The sum of Hats = 45
   SS of Hats = 105.50
   Mean of Gloves = 5.50
   The population variance of Gloves = 12.58
   SSy: = 75.50
   SSxy: = 83.50
   r = .94
   df = 4 (6 people minus 2 equals 4)
   The critical value for r = .81
   Is there a significant relationship between Hats & Gloves at the .05 alpha level? Yes

Day 6: Regression
Making predictions

BRIEFLY

When there is a strong correlation between two variables, you can make accurate predictions from one to the other. If sales and time are highly correlated, you can predict what sales will be in the future...or in the past. You can enhance the sharpness of an image by predicting what greater detail would look like (filling in the spaces between the dots with predicted values).

Regression is an extension of correlation. Now we can count how many rocks you own and use it to estimate how large your hut is.

Regressions allow us to compare data to a straight line and to make predictions about the future and the past. Here are the 4 places our tour takes us today::

Regression
Components
Why regress?
How accurate are our predictions?

INTRODUCTION

Linear Regression

An extension of the correlation, a regression allows you to see how much the data you collect looks like a straight line. Obviously, if your data is cyclical, a straight line won't represent it very well. But if there is a positive or negative trend, a straight line is a good model.

If the data approximates a straight line, you can then use that information to predict what will happen in the future. Predicting the future assumes, of course, that conditions remain the same. The stock market is hard to predict because it keeps changing. Stocks will jump up, decline slowly, head up for a couple of days, and fall down in a matter of hours. It's too erratic to predict its future.

If you roll a bowling ball down a lane and measure the angle it is traveling, you can predict where the ball will hit when it reaches the pins. The size, temperature and shape of the bowling lane are assumed to remain constant for the entire trip, so a linear model would work well with this data. If you use the same ball on a grass lane which has dips and bulges, the conditions are not constant enough to accurately predict its path.

Predicting the future also assumes that the relationship between the two variables is strong. A weak correlation will produce a poor line of prediction. Only strong (positive or negative) correlations will produce accurate predictions.

Linear regression can also predict the past. Carbon dating of a relic assumes that carbon has always burned at the same rate. If we discovered that for some odd reason, carbon burned very quickly 3000 years ago (or very slowly), our predictions of the age of our relic would be substantially off.

Components

A regression is composed of three primary characteristics. Any two of these three can be used to draw a regression line:

First, the regression line always goes through the point where the mean of X and the mean of Y meet. This is reasonable since the best prediction of a variable (knowing nothing else about it) is its mean. Since the mean is a good measure of central tendency (where everyone is hanging out), it is a good measure to use.

If you're asked to guess someone's weight or height (without seeing them), the best guess is the mean of each of those variables. Even if you're wrong, you will typically be less wrong at the mean than any other guess because most of the values in a distribution are at or near the mean.

Second, a regression line has slope. For every change in X, slope will indicate the change in Y. If the correlation between X and Y is perfect, slope will be 1; every time X gets larger by 1, Y will get larger by 1. Slope indicates the rate of change in Y, given a change of 1 in X.

Third, a regression line has a Y intercept: the place where the regression line crosses the Y axis.

Why regression?

We typically use a least-squares criterion to decide where the line should be drawn. That is, the line goes through the data in such a way as to minimize the dispersion from that line.

Regression means to go back to something. We can regress to our childhood; regress out of a building (leave the way we came in). Or regress back to the line of prediction. Instead of looking at the underlying data points, we use the line we've created to make predictions. Instead of relying on real data, we regress to our prediction line.

In a regression, we have moved from the reality of data to a hypothetical straight line. That line of prediction is a representation of what reality might be like. It is not so much that we apply the model to the data; more like we collect the data and ask if it looks like this model (linear), that model (circular or cyclic) or that other model (chance).

Accurate predictions

There are two major determinants of a prediction's accuracy: (a) the amount of variance the predictor shares with the criterion and (b) the amount of dispersion in the criterion.

Taking them in order, if the correlation between the two variables is not strong, it is very difficult to predict from one to the other. In a strong positive correlation, you know that when X is low Y is low. Knowing where one variable is makes it easy to know the general location of the other variable.

A good measure of predictability, therefore, is the coefficient of determination (calculated by squaring r). R-squared (r^2) indicates how much the two variables have in common. If r^2 is close to 1, there is a lot of overlap between the variables and it becomes quite easy to predict one from the other.

Even when the correlation is perfect, however, predictions are limited by the amount of dispersion in the criterion. Think of it this way: if everyone has the same score (or nearly so), it is easy to predict that score, particularly if the variable is correlated with another variable. But if everyone has a different score (lots of dispersion from the mean), guessing the correct value is difficult.

The <u>standard error of estimate</u> (SEE) takes both of these factors into consideration and produces a standard deviation of error around the prediction line. A prediction is presented as plus or minus its SEE.

The true score of a prediction will be within 1 standard error of estimate of the regression line 68% of the time. If the predicted score is 15 (just to pick a number), we're 68% sure that the real score is 15 plus or minus 3 (or whatever the SEE is).

Similarly, we're 96% sure that the real score falls within two standard deviations of the regression line (15 plus or minus 6). And we're 99.9% sure that the real score falls within 3 SEE's of the prediction (15 plus or minus 9).

Regression

Strong relationships allow us to predict behavior. We may not understand causation but with a strong correlation we can make predictions. We can predict into the future, into the past or fill in the gaps between time periods.

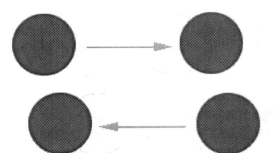

Regression tries to draw a single line through a scatterplot and uses that line to make predictions. Obviously, the more a scatterplot approximates a straight line, the easier it is to draw a line through it. Consequently, regression is only possible with strong correlations, either positive or negative. Correlations of medium to low magnitudes do not lend themselves to regression.

In regression, there is a directionality to the prediction. X may predict Y, or the other way around, but the relationship is not bidirectional. It makes a difference which variable is the predictor and which the criterion. Perspective changes the predictions.

The choice of which variable should be the predictor is based on theory. It is our theoretical questions that determine whether X or Y is the predictor. We use sunset to predict that it will be dark at night, and not the other way around, because of our theory of how the sun functions. If one day the sun goes down and it doesn't get dark, we'll have to dramatically readjust our theory.

To draw a line through a scatterplot of data, we must know where the line begins or a point through which it will go. We also need to know the angle of the line. These two things (a starting point and slope) are essential to drawing a line. The formula for a straight line puts it this way:

$$Y' = a + bX$$

In this formula, "a" is the intercept (the point where the regression line crosses the Y axis), b is the slope (rate of increase) of the line and X is any raw score. Given these three elements, individual points (called Y-primes) can be predicted.

Calculating a regression line begins by finding the **slope** of the line. Here's the formula for the slope of a line:

$$b = \frac{SS_{xy}}{SS_x}$$

Notice that it's a lot like the formula for correlation. The SSxy is on the top and the SSx is on the bottom. What's missing is the SSy, which isn't needed because slope is a measure of commonality to changes in the predictor variable. Let's use this data:

X	Y
2	3
4	5
6	6
8	6
10	8
12	10

The SSx for this data is 70 and the SSxy is 44. To find b, divide 44 by 70. And 44/70 = .629, so b = .63 (rounded to 2 places). In other words, the line of regression moves in a positive direction (in this example b is positive, not negative). And to plot the line, we would move over 1 and up .63 units all along the line. All we need is a starting point.

A common point of interest is where the regression line intercepts the y axis. This intercept is a good visual starting point for drawing a line and can be found if we know the slope of the line and the means of each variable. Here is the formula for intercept:

$$a = \overline{Y} - b\overline{X}$$

Notice that the formula includes slope (which we just calculated) and the mean of each variable. The reason the two means are in this formula is that regression lines always go through the point where the two means meet. This makes sense when you realize that the best predictor of a variable—if you knew nothing else about it—is the mean. Since the mean is our best measure of central tendency and most scores are in the center of a distribution, the mean has the least amount of prediction error.

We calculated the slope to be .63. The mean of Y is 6.33 and the mean of X is 7. So the intercept is 6.33 - (.63 times 7). Working it out, .63 times 7 is 4.40. Then 6.33 - 4.40 equals 1.93. So the intercept of the Y axis (called "a") equals 1.93.

To plot the line, we would go up the Y axis to 1.93 and use that for our standing point. Then we could plot the coordinates of the line by going to the right 1 and up .63, over 1 and up .63, etc.

We don't have to plot the regression line in order to use it. Having calculated what the characteristics the line would have, we can use that information to predict values of Y from values of X.

Using the formula for a straight line, we simply plug in a value for X along with the slope and intercept we've calculated. The X values we plug in need not be known X values. We can extend the line beyond the current data and estimate what Y would be at that point. This extrapolation (estimating beyond the current data set) is very useful for projecting into the future. Budget and sales predictions that are based on previous data are extrapolated into the future.

Predicting inter-points of data is called interpolation. It is possible to estimate midpoints in annual data by interpolation.

Regression assumes that the data is consistent. Future performance is based on past performance, assuming that conditions do not change. Extrapolating sales for next year based on previous years can't take into account a sudden drop in the market or being bought out by a competitor. Interpolating the midyear score on a math achievement test assumes that learning is linear and doesn't happen in spurts.

Clearly, regression works best when conditions are constant and less well in real life. Reality has too many bumps to make precise predictions about health, wealth and happiness. But if you're planning a trip into space and want to know where your ship will be in 3 years (assuming it isn't hit by an asteroid), regression is the right tool.

Regression also assumes that errors of prediction are (a) consistent along the entire regression line and (b) that true values are normally distributed at any given point along that line. A regression assumes that there are an equal amount of errors along the line; error isn't small at one point and larger at another. When a poll says the President's popularity is 38% plus or minus 3, it assumes that those 3 points of error are consistent; the error isn't 3 points at 38 but 5 points at 50%, it's always 3 points. Similarly, a regression assumes that predicting congressional budgets isn't more prone to error at 8 months than it is at 6 months. Error is assumed to be consistent along the line of regression.

At any point along the regression line, prediction forms a normal distribution. The true value of the President's popularity is not known, it is estimated to be 38% but there is some error in that estimate.

However, it is assumed that the true value is likely to be close to the regression line and less and less likely the farther away one gets from that line. Prediction values are assumed to follow a normal distribution, becoming less and less accurate. The regression line is assumed to be the mean of the distribution and that a standard deviation of prediction can be found.

Consequently, predictions along the regression line are made with a standard deviation in mind. The President is at 38% plus or minus 3; sales will be 4 million next year plus or minus 2; you'd have to pay $100 more in income tax next year plus or minus 4. The prediction is stated and the standard deviation follows it.

This standard deviation, called the standard error of estimate, should be interpreted in the same way other standard deviations are used. As a standard deviation, 68% of the scores fall within one standard deviation of the mean, 96% are within two standard deviations of the mean, and virtually all of the scores are within three standard deviations of the mean. With regressions, we're saying that the true score (the real thing when it happens) is likely to fall within one standard deviation of the predicted value.

In other words, we're 68% sure that sales will be between 2 and 6 million next year (4 million plus or minus one standard deviation. We're 96% sure that sales will be between 0 and 8 million (within 2 standard deviations). But we're 99.99% sure that sales next year will be between losing 2 million and making 10 million! Of course, this assumes that nothing changes in the meantime.

As you can see, our margins of error can make our predictions sound silly. We want life to be easy to predict but often it's not. We can make accurate predictions when there is very little error along the regression line. And our predictions become quite vague when the standard error of estimate is large.

There are two things that impact our estimates the most. Both can be seen in the formula for the standard error of estimate:

$$S\,es_{b} = S_{y}\sqrt{\frac{x(1-r^{2})}{n-2}}$$

Prediction errors go up (a) when the standard deviation of Y gets larger and/or (b) when the correlation between the two variables gets closer to zero. First, error increases when the Y's standard deviation increases. That is, it is easier to predict Y when its scores are more homogeneous and it is harder to predict Y when its scores are more dispersed. It makes sense that if everyone has the same score (or close to it), predicting Y is easy. The challenge is to predict Y when its values are dispersed, and the standard error of estimate reflects this difficulty.

Second, the less correlation between the two variables, the harder it is to make accurate predictions. As r gets smaller, the coefficient of nondetermination gets larger. The less variance that can be attributed to the relationship between the two variables, the harder it is to predict values.

UNDERSTAND

Concepts are rules you carry in your head. They are easy to remember and apply to many situations. Here are 3 concepts related to regression.

1. Line of regression

Illustration 1: Regression is a border between countries. It is an imaginary line we use to make decisions.

Illustration 2: Regression is a line on a map. You can draw a straight line from Los Angeles to New York but you have to fly in a curve (the earth is round remember; no fair tunneling through the middle of it). The line is useful for making decisions but is an approximation of reality, not reality itself.

Illustration 3: Regression is like the imaginary line we use to describe where a wall has been built. It is not the wall itself but is a representation of the wall.

2. Consistent errors

Illustration 1: The errors around the regression line are not thick at one end and thin at the other. They are consistent all along the regression.

Illustration 2: The regression line is like a boat with outriggers on each side. There is a center evenly balanced on each side.

3. Three standard deviations of error

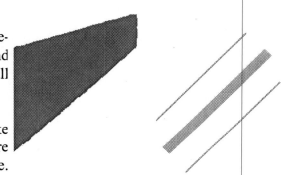

Illustration 1: The errors around the regression line are like a rod through a tunnel of normal curves. Each point on the line has a normal distribution of error surrounding it. The predicted value (dot on the line) is our best guess. We are 68% sure that the real score will be at the dot, plus or minus one standard error of estimate. We are 96% sure that the real score will be within two standard errors of estimate. And we're 99.7% sure that the real score will be within 3 standard errors of estimate.

Illustration 2: The regression line is like a series of guesses surrounded by confidence bands. We're fairly confident that the real score is with one standard error of estimate (SEE) of our predicted score. We're quite sure it will be within 2 SEE and we're really-really sure that it will be within 3 SEE (plus and minus of course).

REMEMBER

Facts are the details of who, what, where and when. Here are the facts I've collected for regressions.

Basic Facts

There are 6 things associated with a linear regression: intercept, slope, interpolate, extrapolate, least square criterion, and standard error of estimate.

Formulas:

$$Y' = a + bX \qquad a = \overline{Y} - b\overline{X} \qquad b = \frac{SS_{xy}}{SS_x}$$

Terms:

a
b
regression line
extrapolation
intercept
interpolation
regression
slope
standard error of estimate (SEE)
straight line
time series
X predicting Y
Y predicting X

DO

Step-by-Step

1 Calculate SSx

2. Calculate SSy

3. Calculate SSxy

3. Find the slope (b)

Divide the SSxy by the SS of the predictor.

$$b = \frac{SS_{xy}}{SS_x}$$

6. Find the Y intercept (a)

Multiply the mean of X by the slope of the regression line. Then subtract the result from the mean of Y.

7. Make a prediction (Y')

Plug the numbers into this formula:

$$Y' = a + bX$$

8. Find standard error of estimate

If you haven't already done so, calculate the Pearson r.

Calculate the standard deviation of Y.

Plug the numbers into the formula for s_{est} (also called SEE or the standard error of estimate):

$$S_{est} = S_y \sqrt{\frac{n(1-r^2)}{n-2}}$$

The result is the standard deviation around the line of prediction

Example

Let's try this example:

X	Y
2	5
4	7
6	6
9	8
12	14
15	11
15	12

First, calculate the mean of X and the mean of Y. Each mean = 9.

Second, calculate the slope of the line (called b). To find the slope, divide the SSxy by the SS of the predictor. That is:

$$b = \frac{SS_{xy}}{SS_x}$$

The SSx is 164; the SSy is 68; and the SSxy 92. So the slope (b) = .56.

Third, calculate the Y intercept (called a). The formula is:

$$a = \overline{Y} - b\overline{X}$$

So a = 9 - (.56 * 9) = 3.95.

Fourth, make a prediction. Use the formula for a straight line:

$$Y' = a + bX$$

Let's assume that the X value is 8, we would predict that Y (which we will call Y-prime so we know it's a prediction) equals 8.44.

Fifth, estimate the accuracy of the prediction. Don't worry, there's a formula for that too. Here it is:

$$S_{est_y} = S_y \sqrt{\frac{n(1-r^2)}{n-2}}$$

The standard deviation of Y (s_y) is 3.12 and r = .87, so the standard error of estimate is 1.81. This means that we're 68% sure the real score will be 8.44 plus or minus 1.81 (that is between 6.63 and 10.25).

Practice Problems

Item 1

Petals	Weight
8	1
5	6
2	7
1	8
9	2

Should weight be X or Y?

a = _____

b = _____

If there are 6 petals, how much will an elephant plant weigh (in pounds):

Item 2

Trucks	Dolls
12	11
8	7
6	5
2	1
4	3
9	9
14	15

Should trucks be X or Y?

a = _____

b = _____

If there are 9 dolls, how many trucks will a child have:

Item 3

Ducks	Cows
3	11
6	6
4	1
15	2
11	10
5	7

a = _____

b = _____

If there are 22 ducks, how many cows will the farmer own:

Item 4

Doctors	Nurses
1	13
11	5
22	3
11	1
16	18
17	8

a = _____

b = _____

If there are 5 nurses, how many doctors will a hospital have:

Item 5

Profs	Classes
4	3
4	8
6	2
8	13
7	8
7	9

a = _____

b = _____

If there are 7 professors, how many classes will be taught:

Item 6

Candles	Flowers
22	2
8	4
3	11
12	4
1	6
6	18

a = _____

b = _____

standard error of estimate (SEE) =

If there are 22 candles, a wedding have _____ flowers.

Item 7

Year	Tuition
1990	1
1991	3
1992	9
1993	12
1994	16

a = _____

b = _____

In 1997 tuition (in thousands) is estimated to be _____

[Hint: use the last digit of the year or convert Years to your own integers]

Simulations

Simulation 1

As a director of marketing, you are interested in how well your sales did over the year. In particular, you are curious whether last year's revenue is a good predictor of this year's. The numbers below are the number of dollars (in millions).

Last Year	This Year $
8	6
13	14
5	4
8	9
11	12
7	11
9	8
2	7

What is the SumX of Last Year:_____

What is the range of Last Year: _____

What is the SS of Last Year: _____

What is the SS of This Year: _____

What is the SSxy: _____

Since you are interested in how well one test acts as a linear predictor of another, which of the following tests should you perform:

    a. t-test
    b. ANOVA
    c. correlation
    d. regression
    e. multiple regression

Perform the comparison you selected in the item above. Select only the appropriate one(s). What was the result of your calculations?\

  a =

  b =

  r =

  t =

  F =

If a client's last year's revenue was 6 million, what would you predict this year's revenue to be?

Simulation 2

Having encountered a new civilization, you use your knowledge of population statistics to make predictions about wealth (number of grey rocks) and housing (hut size).

Rocks	Hut Size
5	11
8	13
2	7
11	9
3	4

What is the SS for Rocks: _____

What is the standard deviation of Rocks: _____

What is the range of Hut Size: _____

What is the SS for Hut Size: _____

What is the SSxy: _____

Since you are interested in how rocks predicted hut size, calculate the following:

a =

b =

If you have 7 rocks, how big is your hut likely to be?

Since you also are interested in how hut size predicts rocks, calculate the following:

a =

b =

If your hut is 17 feet high, how many rocks would you likely have?

Simulation 3

As a politician, you are interested in how family income predicts political contributions. Assume this is a population.

Giving	Income
9	19
13	17
18	22
5	7
8	4
2	4
1	2

What is the SS for Income: _____

What's the pop. stdev. of Income: _____

What is the range of Income: _____

What is the SS for Giving: _____

What is the SSxy: _____

Since you are interested in how well one rating acts as a linear predictor of another, which of the following tests should you perform:

    a.t-test

    b.ANOVA

    c.correlation

    d.regression

    e.multiple regression

Perform the comparison you selected in the item above. Select only the appropriate one(s). What was the result of your calculations?

    $a =$

    $b =$

    $r =$

    $t =$

    $F =$

If a family's income is 4, what would you predict giving to be?

Simulation 4

As a director of counseling, you are interested in how well your therapy works. In particular, you are curious whether the number of counseling sessions is a good predictor of happiness (smiles per hour).

Sessions	Happiness
12	4
3	7
9	2
4	11
7	9
5	5
6	9
2	7

What is the SS for Counseling: _____

What is the pop. stdv. of Counseling: _____

What is the mean of Happiness: _____

What is the SS for Happiness: _____

What is the SSxy: _____

Since you are interested in how well one variable acts as a linear predictor of another, which of the following tests should you perform:

    a.t-test
    b.ANOVA
    c.correlation
    d.regression
    e.multiple regression

Perform the comparison you selected in the item above. Select only the appropriate one(s). What was the result of your calculations?

    a =

    b =

    r =

    t =

    F =

If a client has 10 sessions with you, what would you predict their happiness to be?

SUMMARY

The variable with the smallest standard deviation is the easiest to predict.

Without knowing anything else about a variable, the best predictor of it is its mean. The angle of a regression line is called the slope. Slope is calculated by dividing the S_{xy} by the SS_x.

The point where the regression line crosses the criterion axis is called the intercept.

Predicting the future based on past experience is best done with a regression.

Predicting scores between known values is called interpolation. Predicting scores beyond known values is called extrapolation.

Regression works best when a relationship is strong and linear.

Regression works best when the correlation is strong.

The error around a line of prediction is consistent along the whole line.

The error around a line of prediction can be estimated with the standard error of estimate.

Plus or minus one SEE accounts for 68% of the prediction errors.

A regression is based on paired-observations of the same subjects.

Pre- and Post-test performance is best analyzed by using a regression.

1. Regressions assume that:
   a. Student's t = 0
   b. trends are linear
   c. the variables are discrete
   d. errors vary along the regression line

2. Regressions are best used for measuring:
   a. validity
   b. dispersion
   c. central tendency
   d. distance from a mean

3. The angle of a regression line is called:
   a. intercept
   b. intersect
   c. slope
   d. shift

4. Scores which are projected beyond their samples are said to be:
   a. interpolated
   b. extrapolated
   c. innovated
   d. interpreted

5. Scores which predict between samples are said to be:
   a. interpolated
   b. extrapolated
   c. innovated
   d. unsubstantiated

6. The standard error of estimate is best understood as:
   a. mean
   b. variance
   c. standard deviation
   d. Sum of Squares

7. Linear regressions use the:
   a. central limit theorem
   b. standardized score theorem
   c. peripheral limit theorem
   d. least squares criterion

8. Linear regressions use the formula for a:
   a. circle
   b. curved path
   c. straight line
   d. correlated t-test

9. The line of regression always goes through the point where:
   a. the most error occurs
   b. the two means intersect
   c. the X and Y axes intersect
   d. the slope is balanced

10. As a director of marketing, you are interested in how well your sales did over the year. In particular, you are curious whether last year's revenue is a good predictor of this year's. Since you are interested in how well one test acts as a linear predictor of another, which of the following tests should you perform:
   a. t-test
   b. ANOVA
   c. correlation
   d. regression

Progress Check

1. List three types of correlation and the kind of variables with which they are used:

   a.

   b.

   c.

2. Calculate the following using this data:

X	Y
2	2
3	4
7	8
9	7
11	12
14	10

SSx = _____

SSy = _____

SSxy = _____

a = _____

b = _____

r = _____

3. As a doctor, you are interested in how similar your patients patterns of walking and swimming are. You have measured their performance on each task, and now hope to find how related these two variables are. The numbers below represent the number of hours per week spent by the same patients in each activity.

Walking	Swimming
16	3
11	4
10	4
10	7
5	3
2	12

What is the sum of Walking: _____

What is the mean of Walking: _____

What is the SS for Walking: _____

Since you are interested in communality, which of the following tests should you perform:
a. multiple regression
b. regression
c. correlation
d. t-test
e. ANOVA

Perform the comparison you selected in the item above. What was the result of your calculation (select only the appropriate ones):

a =

b =

r =

t =

F =

How many degrees of freedom are in this study?

What is the critical value for this statistic?

Is there a significant relationship between these variables at the .05 alpha level?

Calculate the coefficient of determination: _____

4. As a chocolate seller, you are interested in how price affects sales. Here is the information from some of the most recent months:

Price	Sales
4	4
5	3
3	7
2	2
9	8
11	12

What is the SS for Price: _____

What is the variance of Price: _____

What is the range of Price: _____

What is the SS for Sales: _____

What is the SSxy: _____

Since you are interested in how well one rating acts as a linear predictor of another, which of the following tests should you perform:

        a.t-test
        b.ANOVA
        c.correlation
        d.regression
        e.multiple regression

Perform the comparison you selected in the item above. Select only the appropriate one(s). What was the result of your calculations?

$a =$

$b =$

$r =$

$t =$

$F =$

If price equals 15, what would you predict sales to be?

Answers

Practice Problems

Item 1
Petals (X) is predicting Weight (Y)
SSx = 50; SSy = 38.80
SSxy = - 42; a = 9; b = -.84
If Petals is 6, predict weight = 3.96

Item 2
Dolls is X, Trucks is Y
SSx = 139.43; SSy = 108.86
SSxy = 122.29; a = 1.47; b = .88
If Dolls = 9, predict Trucks = 9.39

Item 3
SSx = 109.33
SSxy = - 23.33; a = 7.73; b = - .21
If have 22 ducks, predict cows = 3.11

Item 4
SSxy = - 55; a = 15.12; b = - .26
If 5 nurses, predict doctors = 13.82

Item 5
SSxy = 21; a = - 1.83; b = 1.50
If 7 professors, predict 8.67 classes

Item 6
SSxy = -119; a = 11.09; b = - .41
If 22 candles, predicts 2.07 flowers.
SEE = 5.71

Item 7
SSx = 10; SSy = 154.80
SSxy = 39; a = .40; b = 3.90
Predict tuition to be 27.70 in 1997

Simulations

Simulation 1
ΣX = 63; range of X = 11
SSx = 80.88; SSy = 76.88
SSxy = 57.88
Regression
a = 3.24; b = .72
If last year = 6, this year = 7.56

Simulation 2
SS of Rocks = 54.80
Population stdev = 3.31
Range of Huts = 9
SS of Huts = 48.80
Rocks predicts huts
  a = 5.75
  b = .53
  If 7 rocks, 9.43 huts
Huts predicts rocks
  a = .61
  b = .59
  If hut 17 feet high, 10.64 rocks

Simulation 3
SS of Income = 415.43
Population stdev of Income = 7.70
Range of Income = 20
SS of Giving = 220
SSxy = 265
a = 1.17
b = .64
If family income = 4, giving = 3.72

Simulation 4
Counseling
  SS= 76
  Population stdev = 3.08
Happiness
  Mean = 6.75
  SS = 61.50
SSxy = - 37
a = 9.67
b = - .49
If 10 sessions, happiness = 4.80

Multiple Choice
1. b, 2. a, 3. c, 4. b, 5. a,
6. c, 7. d, 8. c, 9. b, 10. d

Progress Check
1. List three types of correlation and the kind of variables with which they are used:
  a. phi   2 discrete variables
  b. Pearson r   2 continuous variables
  c. point-biserial 1 continuous &
                    1 discrete variable

2. Calculate the following using this data:
   SSx = 107.33
   SSy = 68.83
   SSxy = 77.33
   a = 1.67
   b = .72
   r = .90

3. Walking & Swimming
   Sum of Walking: 54
   Mean of Walking: 9
   SS for Walking: 120
   Correlation
   r = -.65
   degrees of freedom   4
   Critical value   .811
   Not significant relationship at the .05 alpha level
   Coefficient of determination: .43

4. Chocolate Price & Sales
   SS of Price: 63.33
   variance of Price: 12.67 (sample variance for "some" of the months)
   range of Price: 9
   SS for Sales: 70
   SSxy: 56
   regression
   Price is X, Sales is Y
   a = 1.01
   b = .88
   If price equals 15, what would you predict sales to be?   14.21

Day 7:
Probability
Comparing a group to a standard

BRIEFLY

Today, we're going to take a quick tour of the Isle of Chance. It's a place where everything is random. Occupied by odds and probabilities, the whole island is a haven for stray thoughts and quirky behaviors.

Have you ever had an amazing coincidence? Seen someone you know coming off a plane while you're waiting to get on that same plane? Gone to a concert or football game and run into a friend you didn't know would be there? Traveled to a far off country only to find people from your home town there? Then, you've been to the Isle of Chance.

Chance is where you'll find probabilities, odds, estimation, and, of course, gambling.

It's also where mathematics and philosophy often collide. Did the universe begin by chance? What are the odds there is life after death? What's the probability that we have free will, unique personalities or extra-sensory perception? These are among the many questions I won't answer. I don't know the answers. But you get to answer all of them, some of them or none of them. This is where your personal philosophy, religion and culture come into play. How much is chance a part of life? Let me describe it and you decide.

INTRODUCTION

Chance

Philosophically, we live in a world of likelihoods. We know that it rains in the Sahara but that it doesn't rain frequently. We know that snow in January is likely in New York and unlikely in Tahiti. We know that heavy objects fall. Notice that we have no proof that the next time we drop a bowling ball that it will fall. It's just that we think it is highly likely. We think it is highly likely because we don't know of any time that it hasn't occurred that way. Imagine all of the theories we'd have to change if we woke up tomorrow and every time you dropped something it went up.

In social science, business and real life, we base many of our decisions on probabilities. We believe elevators will probably take us to the correct floor, planes will probably fly and our job will probably be there tomorrow. Notice we're not saying that our prediction forces something to happen. We are simply making guesses, predictions and estimations of the likelihood of events.

Probabilities & Odds

If you take a die (one of a pair of dice) and roll it, what is the probability of rolling a 3?

The answer depends on how many 3s are on the die. If you used a die with a 3 on each side, the probability is 1. If you used a die without a 3 on any side, the probability is 0. Probabilities range from 1 (100% certain) to 0 (completely impossible). Anything in between 1 and 0 is expressed as a decimal (eg. .9, .67, .05, etc.).

If you were using your ordinary, friendly six-sided die with the numbers 1 through 6 (one on each side), the probability of rolling a 3 is .167. This is also the probability of rolling any of the other numbers. Each number has an equal chance; each has a probability of .167.

Probabilities are comparisons with the total possible. The probability of picking an ace out of a deck of 52 cards is the number of aces in the deck (4) divided by the total number of cards (52). The result is expressed as p = .077.

Although thought to be interchangeable terms, odds are different from probabilities. Odds are descriptions of for and against. The odds of picking an ace out of a deck of 52 cards is the number of aces in the deck (4) to the number of cards that aren't aces (48). So the odds are 1:12 in your favor (if you're optimistic) or 12:1 against you (if you're pessimistic).

By convention, odds are described in ratios and probabilities are described as decimals. The odds of choosing the correct answer on a multiple choice test with four options is 1:3 or three to one against you. The probability of choosing the correct answer on a multiple choice test is 1 out of 4 or .25. Similarly, the odds of getting heads when flipping a coin is 1:1 (1 head and 1 tail). But the probability of getting heads would be 1 out 2 or .50

Flipping a coin only has two alternatives so it's easy to predict the result. It will be either heads or tails. One outcome out of two total possibilities. So the probability is 1/2 or .50. Similarly, the probability of winning a raffle depends on the number of tickets you buy divided by the total number of tickets sold. If there are only 10 tickets and you buy 1, the probability is 1 divided by 10 or .10. If there are 1000 tickets, the probability drops to .001.

But what is the probability of winning 2 separate lotteries?

Calculating multiple events

To calculate the probability of two independent events occurring at the same time, we multiply the probabilities. If raffle #1 has 10 tickets and raffle #2 has 1000 tickets (and you only buy 1 ticket of each), the probability of winning both raffles is .1 times .001. The product of the probabilities equals .0001.

Similarly, if the probability of you eating ice cream is .30 (you spend a lot of time eating ice cream) and the probability of your getting hit by a car is .50 (you live in the middle of a road), the probability that you'll be eating ice cream when you get hit by a car is .15. Flipping a coin twice (2 independent events) is calculated by multiplying .5 times .5. The probability of rolling 2 heads in a row is .25. Rolling snake eyes (ones) on a single roll of a pair of dice has a probability of .03 (.167 times .167). To calculate multiple independent events, multiply the simple probabilities of each accruing by itself.

Sometimes it's easier to calculate the chances of things not occurring and subtract it from 1 (the total possible probability of things occurring and not occurring). A good example is calculating the probability that strangers in a room will share the same birthday. It is easier if we calculate the probability that none of them share the same birthday and subtract that number from one.

Let's start with you in a room. The probability of your having your birthday is 1. Also the probability of another person in the room not having your birthday is 1 (remember, you're the only one in the room, there is no one else to match).

When we add one person (Anna), the probability of that person not having your birthday is 364/365. Expressed as a probability, p = .997. So the probability of Anna sharing your birthday is 1-.997 or .003.

When we add another person, Bob, the pool of available days is smaller. Anna's birthday is taken. So the probability for Bob not sharing a birthday with either you or Anna is Anna (364/365) times Bob (364/365), or .995. Subtracting this from 1, and we find the probability is .005. Naturally, as we keep adding people, the probability of 2 people sharing the same birthday increases.

It doesn't take that many people for coincidences to happen. In a room of 23 people, there is a 50:50 chance that at least two of them will have the same birthday. With 40 people, the probability is .89 and with 50 people it's .97.

Notice this is not the probability of them having your birthday. It takes 253 people to have a 50:50 chance that one of them will share your birthday. And that's an important distinction. The probability of things happening in general is not the same as the probability of them happening to you.

The probability of things happening to us is less than the probability that they will happen in general. And that's sort of how we think but we overestimate the unusualness of things in our lives. We don't think other people's coincidences are as amazing as ours. If they are surprised to see their uncle in France, we are overwhelmingly flabbergasted. Other people's coincidences are interesting but ours are amazing. We are biased by our personal involvement. It's not surprising that our estimates are off. We are much more involved in our lives than in others. What happens to us is, after all, special simply because it happens to us.

We are biased by our experience too. If I told you that last month all of the babies born in a local hospital were boys, your knowledge of hospitals might come into play. If you're used to large hospitals, you might think this a very unusual situation. But if you're more familiar with very small hospitals, you might not be surprised at all. If it was a tiny hospital where only one baby was born, it would not be surprising that all of the babies were male. If it was a large hospital with dozens of babies born, it's extremely unlikely that all of the babies would be boys.

We can't experience everything, so the list of things we've encountered is rather small. Most of our decisions are based on very small samples of behavior. In fact, the smaller the sample, the more we rely on it.

I've met several members of the U.S. House of Representatives. I've never personally met any U.S. Senators but I've seen one President in person. My understanding of Representatives has some variety in it. If I met another one who was a bit odd, I could adjust my conclusions about people in Congress slightly to account for those actions. But all I know about how Presidents act in person is based on one person. If he was weird, I would think all Presidents were weird and it would take meeting several more to convince me that my original conclusions were wrong. The smaller the data pool, the more influence it has on us.

We're also biased in our estimates because we rely on exceptions. If we're told that smoking cigars cuts your life short, we think of a really old person who smokes cigars every day. If we're told that the type of car we own was shabbily built, we point to the one we own as an exception. We tend to believe that exceptions disprove rules. But general rules are just that: general. They are broad descriptions of group data. A new wonder drug might cure the common cold for most people but not work on you. You don't disprove the rule; you simply aren't covered by it.

The broader explanation for these biases is our belief in the Law of Small Numbers. We tend to think that rules based on a few examples will extend to a larger sample. If we start with the rule that every odd number is a prime number, it seems pretty good at first. After all, 1 is a prime number. So is 3. The rule holds for 5 and 7. If we stop here, we might conclude that the rule will continue on forever. But 9 is not a prime number and the rule fails after only 5 trials.

This tendency to believe in rules can be seen in gambling. The "gambler's fallacy" is the tendency for us to think that an event is more likely to occur because it hasn't happened for a long time. If we're throwing a die and only odd numbers come up, we think it's more likely that the next throw will result in an even number. If a roulette wheel hits the same number 4 times in row, it doesn't mean that it can't hit the same number another 4 times in a row.

If you flip a coin 20 times and every time it comes up heads, the probability it will be tails next time is .5, the same as always. Coins have no memory. They fly through the air, land on their side, and have no recollection of the event.

Some people try to double their bets every time they lose. They figure they will eventually win back all of their losses and more. This approach will work every time ONLY if you have an infinite amount of money and an infinite amount of time. You have to be able to continue doubling your bets long after you've left Vegas. Usually, doubling your bet just means that you lose faster.

Chance has a pattern but it's a very odd pattern. We're much more organized than chance. If you have a lot of friends, try this. Ask half of the people in a large group to flip a coin and record how many heads came up. And ask the other half to mentally flip a coin and record how many heads came up. The mentally flipped coin sequences usually underestimate the length of a lucky streak. Over a course of 250 tosses or so, about a third of them will have at least 8 heads.

One of chance's odd patterns is that the number 1 comes up more often than you'd predict. Take a large list of numbers for stock prices, street addresses, lengths of rivers, etc. and see what the first digit is. Digits vary from 0 to 9 but they are not equally used. The number 1 comes up about 30% of the time. This "first-digit law" (also called Benford's law) is underestimated by people. When embezzlers forge financial records, the numbers they generate don't begin with 1 as often as real data does.

I like talking about probability at this point in the tour because we've been using it all along but without much fanfare. When we looked at the normal curve and saw that most people are in the middle of the distribution, we were in effect saying that the probability of being in the middle is higher than being at either end.

We used z-scores to set criterion at the 95th percentile. We were saying that scoring at about the 95th percentile was unlikely. The probability of scoring above our set z-score was .05 (5%). When we tested the significance of the correlation coefficient, we used probability. When we predicted a Y' from a regression line, we were using probability. Every time we test for significance, we are making statements of probability.

Probability is not foreign to what we're doing; it's built in. Notice how much we rely on probability as we explore a new technique: Analysis of Regression.

Analysis of Regression

Typically, we collect data and compare it to a model. We ask "Does the data look like this?" "This" can be a simple or complex model. Nearly all group statistics are trying to match data to a model. The model differs from procedure to procedure but the underlying premise is the same. Let's take a simple model and extend our understanding of the linear regression to model testing. You already know how to calculate a correlation and a regression, so let's expand on that knowledge.

A correlation is a measure of commonality; how much two variables have in common. For the Pearson r, we plotted two continuous variables and looked at the scatterplot of the data. We could see if the trend was generally positive, negative, or had no linear pattern.

A regression is used from predicting. We plot a regression line through the data as best we can, using the line to make predictions. We can predict into the future (extrapolation) or fill in between data points (interpolation).

An analysis of regression looks at the pattern of data and compares it to the regression line drawn through it. It asks how well the data looks like a straight line. This is a yes-no comparison. We start with the premise that the data doesn't look like a straight line. We assume that there is no pattern. When we see small variations from a chance pattern, we still don't accept the model of a straight line. We only change our minds when the pattern is so strong that it is significant.

Significance

Our test of significance is actually a ratio of knowledge. The F test (named after its author, R.A. Fischer) is the ratio of understood variance (called <u>regression mean squares</u> or mean squares regression) to non-understood variance (called <u>error mean squares</u> or mean squares error).

After you calculate the F, you compare it to the critical value in a table of <u>Critical Values of F.</u> In some statistics books there are several pages of critical values to choose from because the shape of F distribution changes as the number of subjects in the study decreases. Table D at the end of this book gives a simplified version of the values but the process is the same. To find the right critical value, go across the degrees of freedom for regression (df regression) and down the df error.

<u>Degrees of freedom</u> (df) are the number of items that are free to varying, assuming a fixed mean. Usually, we start with a few scores (2, 3,7) and try to find the mean (4); we begin with the numbers being fixed and the mean free to be calculated. But degrees of freedom requires backward thinking. We start with the mean and try to figure out the numbers used. If we begin knowing that the mean is 4, it seems like the three numbers in our data set could be any three numbers. But not quite.

Let's assume the first number is free to vary; it can be as large or small as you wish. And let's assume that the second number is free to vary; again it is entirely your choice. But having selected these two numbers, there is only one number that can be used in combination with the other two numbers to produce a mean of 4.

If the first number was 6 and the second number was 2, the third number has to be 4 in order for the mean of the data set to be 4. If the first number was 100 and the second number was 2, the third number has to be some big negative number. It might take me a bit of time to figure out that it's -90, but the third number has got to be that number. It is not free to vary. Degrees of freedom is like a parlor trick that starts with a known mean and allows all the numbers to vary except one (or sometimes a few). Each Sum of Squares in our summary table has its own degrees of freedom.

Using the df's of regression and error, we enter the table to find the right critical value to use as our standard. Go across the degrees of freedom for regression (df regression) and down the df error. There you will find the critical value that F has to beat in order to be significant. Notice that our table only has values for alpha levels of .05 because typically the standard we use for significance only allows 5% error. That is, significance is granted only when we will be right 95% of the time, and choose wrong 5% of the time. In other words, our <u>alpha level</u> is set at .05 (the amount of error we are willing to accept). Setting the criterion at .05 alpha indicates that we want to be wrong no more than 5% of the time. Being wrong in this context means to see a significant relationship where none exists.

Two points should be made: (a) 5% is a lot of error and (b) seeing things that don't exist is not good. Five-percent of the population of the US (if it had 300 million people) is 15 million people; that's a lot of error. If elevators failed 5% of the time, no one would ride them. If the OPEC oil cartel trims production by 5%, they cut 1.5 million barrels a day. There are 230 million people who use the internet, about 5% of the world's population. Cars, trains, planes and elevators would be considered totally unsafe if 5% of them crashed every day. It might seem odd that we'd complain if 1 out of 20 donuts was rotten, but we're fine when 1 out of 20 experiments finds results that aren't true. But we expect more from our donuts than we do from people.

We use a relative low standard of 5% because our numbers are fuzzy. Social science research is like watching a tennis game through blurry glasses which haven't been washed in months. We have some understanding of what is going on—better than if we hadn't attended the match—but no easy way to summarize the experience.

Second, seeing things that don't exist is dangerous. In statistics, it is the equivalent of a hallucination. We want to see the relationships that exist and not see additional ones that live only in our heads. Decisions which produce conclusions of relationship that don't exist are called <u>Type I</u> errors.

If Type I error is statistical hallucination, <u>Type II</u> error is statistical blindness. It is NOT seeing relationships when they do exist. Not being able to see well is the pits (I can tell you from personal experience) but it's not as bad as hallucinating. So we put most of our focus on limiting Type I error.

We pick an alpha level (how much Type I error we are willing to accept) and look up its respective critical value. If the F we calculate is smaller than the critical value, we assume the pattern we see is due to chance. And we continue to assume that it is caused by chance until it is so clear, so distinct, so accurate that it can't be ignored. We only accept patterns that are significantly different from chance.

Fortunately, we don't allow chance to be on our side and there are only two types of decision errors we can make. We set up our studies so that chance is our primary explanation. Unless there is substantial evidence to the contrary, we assume that the relationships between variables are due to chance.

We phrase our hypotheses in terms of null findings. We assume there is no relationship between variables, no difference between groups, and no significant impact of one variable on another. Unless we have evidence to the contrary, we <u>accept the null</u> hypothesis. We need a significant amount of evidence for us to <u>reject the null</u> hypothesis and to say that there is a significant relationship between variables.

This approach is equivalent to the presumed innocence of a criminal suspect. People are assumed to be innocent until proved guilty. And when they are not proved guilty, they are not innocent, they are not guilty. Similarly, the lack of a significant finding doesn't mean that a causal relationship doesn't exist, only that we didn't find it. This is particularly true of the F test, which is a 1-tailed test of significance. We are testing whether the F ratio of explained to unexplained variance is significantly different from what we would expect by chance. If F is not significant, we can make no negative causal statement (i.e., that X does not cause Y). If F is not significant, the decision is that the relationship is likely to be due to chance, not that it is due to chance.

When the F we calculate is larger than the critical value, we are 95% sure that the pattern we see is not caused by chance. By setting the alpha level at .05, we have set the amount of Type I decision error at 5%. When we inappropriately reject the null (find people guilty who aren't), we are making a Type I decision error. Notice that it isn't the data that is incorrect, it is the decision that is in error. The responsibility for the decision lies with the researcher, not the data.

If we accept the null when we shouldn't (find people innocent who aren't), we are making a Type II decision error. Just as in the judicial system, we dislike making Type II errors but we believe that it is better to err on the side of caution.

In an F test, we set the alpha level (the amount of errors we are willing to accept) to .05. That is, we are willing to be wrong 5 times out of 100. If we select .10 or .15 or .20 alpha levels, we would be choosing to allow more Type I error. If we select a .01 alpha level, we would have less Type I error. Most researchers choose an alpha level of .05; it is the standard level of acceptable error.

Type II error has less to do with where we set the alpha level than with the quality of the test itself. Usually we make Type II errors because our instruments are not refined enough to detect significant changes in variables. We don't find significant results because our tests use ordinal information far more than we would like to admit.

The F test is a one-tailed test of significance. The only choices are one- or two-tailed tests. In a one-tailed test, the entire alpha range (all 5%) is at one end of the distribution (at one tail). One-tailed tests answer the question of whether or not something is significantly better than the rest of the distribution. You'd use such a test to see if the magic drug you created is significantly better than the current treatment. Notice, the test says nothing about whether the new drug is significantly worse.

In a two-tailed test, the alpha range is split between both ends of the distribution. That is, an alpha of .05 has two regions of significance, each composed of 2.5%. Two-tailed tests answer the question of whether your new drug is significantly better or significantly worse than the current treatment.

An Analysis of Regression tests one tail: to see if there is significance. In this case, it checks to see if X has a significant impact on Y. It says nothing about Y's impacting X. It only asks if the data differs significantly from chance.

Getting from SS to ms

The F test is a ratio of variance (understood/not understood). To find the variance, we begin by partitioning the Sum of Squares (SS) of the regression into explained and unexplained components. Explained variance is simply the SSy multiplied by r^2 (the coefficient of determination). The result is the SSregression (the understood portion of the regression).

The unexplained, not yet understood portion of the regression is found by multiplying the SSy by $1-r^2$ (the coefficient of nondetermination). The result is the SSerror (the non-understood portion of the regression).

To get from Sum of Squares to variance, we divided each SS by its respective degrees of freedom. The resulting variance terms are called mean squares (a reminder that variance is the average of the squared deviations from a distribution's mean).

The degrees of freedom (df) for Regression is k-1 (columns minus one). Since a simple linear regression has only 2 columns, the df for an Analysis of Regression always equals 1. The df for Error is N-k (number of people minus the number of columns). And Total error = N-1.

If it seems like it's getting hard to keep track of all this, there is good news. An Analysis of Regression uses a summary table that organizes all of the important information. Simply fill in the blanks of the table and the hard part is done.

Example:

X	Y
2	4
5	7
3	9
6	8
11	10
12	10

In this example, the SSx is 85.5, the SSy is 26 and the SSxy is 36. The correlation between the two variables equals .76.

In an Analysis of Regression, the SStotal equals the SSy; in this example it equals 26. To partition this into the portion explained by the regression, multiply SStotal by r-squared (r^2). In this example, it is 26 times .58, which equals 15.08.

The SSerror is the SStotal times the coefficient of nondetermination ($1-r^2$); in this case that would be .42 times 26 = 10.92. Of course you also could subtract the SSregression from the SStotal. Either way will work.

So far the summary table would look like this:

	Sum of Squares	df	mean squares
SS regression	15.08		
SSerror	10.92		
SStotal	26		

We've partitioned the Sum of Squares into the portion explained by the regression (15.20) and the portion that is due to error (10.92). In this context, anything that isn't explained by the regression line is considered error.

Our summary table now looks like this:

	Sum of Squares	df	mean squares
SS regression	15.08	1	
SSerror	10.92	4	
SStotal	26	5	

Mean squares is another name for variance. As you will recall, variance for a single variable is SS divided by N (it's degrees of freedom). So each of the SS is divided by its degrees of freedom. The results are shown here:

	Sum of Squares	df	mean squares
SS regression	15.08	1	15.08
SSerror	10.92	4	2.73
SStotal	26	5	5.20

Of course, the mean squares won't add up like the other columns because we divided by different amounts But the resulting variance terms are appropriate for their respective portions.

The mean squares are compared by dividing the mean squares of regression by the mean squares of error. The result is called F. In this example F = 5.52. In order for the value we calculated to be deemed significant, if must be larger than a standard value for that size of a data set.

We compare the F we calculated to the F table at the back of nearly any statistics book. To find the right value, we select the first column (the same value as the df for SSregression). And to find the correct row, we go down to the row labeled 4 (the same value as the df for SSerror). In this case the book value is 7.71. Our F value was 5.52 so we lose.

Well, lose isn't really the right word but that's always how it seems to me. I think of it as trying to beat the book value and if our F is larger than the book's, we win. If our F is smaller than the book's, we lose.

The proper explanation is that F indicates the likelihood that what we see is not due to chance. If our F is smaller than the book, what we see is likely to be due to chance. If our F is larger than the book, the relationship between variables is likely to be due to something other than chance.

The F test doesn't tell us what causes what, only whether it is a likely occurrence or not. In this example, there is no significant impact of X on Y. Any apparent causal relationship can be explained by chance.

The F test, and other tests of significance, start with the assumption that the relationships we find are due to chance. This presumption is maintained until we are highly confident that it is not true. We want to be very careful that we don't see relationships that don't exist (the statistical equivalent to being psychotic).

We're not as upset about not seeing relationships that do exist (being blind) because we figure that if we replicate the findings enough we will eventually discover their true nature. We may not be fond of being blind (called Type II error) but we hate the idea of being psychotic (called Type I error).

UNDERSTAND

Sampling with and without refreshment

After we've done a card trick, the odds of selecting a particular card for the next trick depends on whether we put the previously used cards. It matters if we base our calculations on cards that have been refreshed (put back) or if our sample of cards was selected without refreshment.

Illustration 1

Sampling with refreshment is like taking a cookie from the cookie jar, licking it and sticking it back in the jar. Whoever comes along next, including you, has all of the original choices you had.

Illustration 2

Sampling without refreshment is like taking a cookie from the cookie jar and eating it. Once gone, nobody, including you, can choose that cookie again.

Goodness of fit

We use Analysis of Regression to test the "goodness of fit" of our data. A data set that fits well approximates a straight line. Data sets that are less linear do not have goodness of fit.

Illustration 1

Goodness of fit is like the way you shop for shoes. You check to see if your new shows fit better than your old ones. Or you check to see if your new shows are more in fashion than your previous pair. You want to see if your shoes vary from the standard form (your current comfort or current style).

Illustration 2

Goodness of fit is like the way you shop for jeans. You check to see if your jeans fit your body. You want to see if your clothes vary from the standard form (your body). You could set the null hypothesis to be "no different from current jeans." Then you'd conduct a test (wear them and measure your sensations and other people's reactions). After testing, you'd either accept the null hypothesis (no better than before) or reject the null hypothesis and accept the alternative hypothesis (you look great in these jeans...significantly better than before).

REMEMBER

Basic Facts:

Theories are composed of constructs; models are composed of variables. Laws have probabilities whose accuracy is beyond doubt. Principles have some predictability but the probability of beliefs is a matter of personal opinion.

Formulas:

An Analysis of Regression uses the formulas for correlation and a summary table which uses the following formulas:

$$\text{Regression}_{df} = k-1 \qquad \text{Error}_{df} = N-k$$
$$\text{Total}_{df} = N-1 \qquad \text{Mean sqaures} = SS/df$$
$$F = \text{Regression}_{df}/\text{Error}_{df}$$

Terms:

alpha level
Analysis of Regression
checklist
criterion reference testing
critical values of F
df
F
k
k-1
mean squares
mean squares$_{error}$
mean squares$_{regression}$
N-1
N-k
partitioning
regression mean squares
Type I error
Type II error

DO

Step-by-Step

X	Y
22	15
5	3
4.50	6
8	10
5.50	1

1. Calculate the Pearson r

$SSx = 218.50$ $SSy = 126$ $SSxy = 142.50$
$r = .86$ $r^2 = .74$

2. Make a summary table

In order to test the significance of a linear regression, start by creating the following table:

	SS	df	ms
Regression	____	___	____
Error	____	___	____
Total	____	___	____

3. Partition SSy

In order to test the significance of a linear regression, the SSy is partitioned into two parts: the amount due to the regression and the amount due to chance.

First, enter the total SS (Sum of Squares) for the table, which is the SSxy.

Second, multiply the SSy by r^2.

Since the coefficient of determination (r^2) is a measure of variance accounted for, multiply SSy by it will give the percentage of SS due to the regression.

Third, multiply the SSy by $1 - r^2$.

The coefficient of nondetermination ($1 - r^2$) times SSy indicates the percentage of SS due to error (unaccounted for variance; what we don't know).

Updating the table, we now can see:

	SS	df	ms
Regression	93.24	___	____
Error	32.76	___	____
Total	126	___	____

4. Find degrees of freedom

First, enter the degrees of freedom (df) for Regression, which is k-1 (columns minus one). Since a simple linear regression has only 2 columns, the df for regression = 1.

Second, enter the df for Error, which is N-k (number of people minus the number of columns). In our example, $N = 5$, so $df_{error} = 3$.

Third, enter the df for Total, which N-1 (number of people minus one). So, $df_{total} = 4$. Updating the table, we now can see:

	SS	df	ms
Regression	93.24	1	____
Error	32.76	3	____
Total	126	4	____

In order to check the accuracy of your calculation, simply add the dfs together; Regression + Error should equal Total.

5. Find mean squares

Mean squares is another name for variance. And since SS divided by df equals variance, divide each SS by its respective df.

Updating the table, we now can see:

	SS	df	ms
Regression	93.24	1	93.24
Error	32.76	3	10.92
Total	126	4	31.50

6. Calculate F

Our test of significance is called an F-test. It is a ratio of the variance we understand to the variance we don't understand.

	SS	df	ms
Regression	93.24	1	93.24
Error	32.76	3	10.92
Total	126	4	31.50

Simply take the mean squares for Regression ($ms_{regression}$) and divide it by the ms_{error}. The result is F. So, F = 8.54

7. Find the critical value

The Critical Values of the F Distribution table is actually a series of distributions. To enter the table, go across to the row whose number matches the degrees of freedom for Regression ($df_{regression}$). And go down the df_{error}.

In our example, go across to 1 and down to 3. The critical value (the value you have to beat) = 10.13 (at the .05 alpha level).

8. Test significance

To test the significance of a linear regression, compare the F you calculated with the critical value found in the table. If your F is larger than the book, you win: the regress is significant and the relationship is not likely to be due to chance.

In our example, we calculated F to be 8.43. The critical value in the F table was 10.13, so F is not significant. And the relationship between the two variables is likely to be due to chance.

Practice Problems

Item 1

Calculate an Analysis of Regression on the following data, where Reading is predicting Education:

Reading	Education
15	11
7	8
2	5
9	4
3	4
11	9

What is the Pearson r:

Perform the following calculations:

	SS	df	ms
Regression	____	____	____
Error	____	____	____
TOTAL	____	____	

What is the F for this test:

Item 2

Calculate an Analysis of Regression on the following data, where TV watching predicts Violence:

Television	Violence
2	7
4	14
8	12
13	8
16	5
19	1

Perform the following calculations:

	SS	df	ms
Regression	____	____	____
Error	____	____	____
TOTAL	____	____	

What is the F for this test:

Item 3

To discover if TV watching has a significant impact on people's shyness, use the following information, perform an Analysis of Regression and complete the following summary table:

$r = -.55$
$SS_x = 20.22$
$SS_y = 15$
$N = 22$

	SS	df	ms
Regression	____	____	____
Error	____	____	____
TOTAL	____	____	

What is the F for this test: _____

Is the critical value for F: _____

Is the F significant?

Item 4

To discover if learning statistics has a significant impact on people's friendliness, use the following information, perform an Analysis of Regression and complete the following summary table:

$r = .45$
$SSx = 17.33$
$SSy = 16.30$
$N = 18$

	SS	df	ms
Regression	___	___	___
Error	___	___	___
TOTAL	___	___	

What is the F for this test: _____

What is the critical value for F: _____

Item 5

To discover if education has a significant impact on people's income, use the following information, perform an Analysis of Regression and complete the following summary table:

$r = .73$
$SSx = 50$
$SSy = 74$
$N = 12$

	SS	df	ms
Regression	___	___	___
Error	___	___	___
TOTAL	___	___	

What is the F for this test: _____

What is the critical value for F: _____

Is the F significant?

Item 6

To discover if music lessons have a significant impact on people's intelligence, use the following information, perform an Analysis of Regression and complete the following summary table:

$r = .12$
$SS_x = 40$
$SS_y = 42$
$N = 13$

	SS	df	ms
Regression	____	____	____
Error	____	____	____
TOTAL	____	____	

What is the F for this test: _____

What is the critical value for F: _____

Is the F significant?

Simulations

Simulation 1

As president of associated students, you wonder how much influence alumni giving has on the amount of scholarship money distributed. The numbers below are the number of dollars (in thousands) for some of the last few years:

Alumni	Scholarships
13	12
9	7
6	5
7	3
4	7
1	2

What is the SS for Alumni: _____

What is the variance of Alumni:_____

What is the range of Alumni: _____

What is the SS for Scholarships:_____

What is the SSxy: _____

Perform the following calculations:

	SS	df	ms
Regression	____	____	____
Error	____	____	____
TOTAL	____	____	

What is the F for this test:

Is it significant at .05 alpha?

Does alumni giving have a significant impact on scholarship money?

Simulation 2

As a Realtor, you wonder how much influence Age (in decades) has on selling prices of Houses (in $50,000s).

Age	House
10	11
3	2
5	5
7	9
9	8
2	1

What is the SS for Age: _____

What's the sample variance of Age:_____

What is the range of Age: _____

What is the SS for House: _____

What is the SSxy: _____

Perform the following calculations:

	SS	df	ms
Regression	____	____	____
Error	____	____	____
TOTAL	____	____	

What is the F for this test:

Does age have a significant impact on house price?

SUMMARY

The probability of X and Y occurring at the same time equals the probability of X times the probability of Y.

The probability of X or Y occurring (either one) is the addition of the probability of X and the probability of Y.

Analysis of Regression tests the likelihood that the linear relationship between the two variables is due to chance. A significant F indicates that X predicts Y well and that the relationship between the two variables is not likely to be due to chance.

Analysis of Regression does not prove cause and effect. X may cause Y or Y may cause X, or both could be caused by another variable.

1. If the probability of being hit by lightning is .3 (it's a very stormy night), and the probability of eating chocolate ice cream is .5, what is the probability of eating ice cream AND being hit by lightning?
    a. .3
    b. .5
    c. .8
    d. .15

2. In the simplest case, the probability of either A or B occurring is calculated by:
    a. adding the probabilities
    b. subtracting the probabilities
    c. multiplying the probabilities
    d. dividing the probabilities

3. If the alpha level is set at .05, the probability of making a Type I error is:
    a. .01
    b. .02
    c. .05
    d. .10

4. The odds of guessing the right answer of this item by chance are:
    a. 1:2
    b. 1:3
    c. 1:4
    d. 2:3

5. A test of significance is said to be:
    a. no-tailed
    b. one-tailed
    c. two-tailed
    d. three-tailed

6. The probability of rejecting the null when you shouldn't is called:
    a. Type I error
    b. Type II error
    c. Type III error
    d. Type O error

7. The probability of accepting the null when you shouldn't is called:
    a. Type I error
    b. Type II error
    c. Type III error
    d Not My Type error

8. An Analysis of Regression tests a regression's:
    a. reliability
    b. validity
    c. symmetry
    d. goodness of fit

9. An Analysis of Regression uses a:
    a. b test
    b. t test
    c. F test
    d. vocabulary test

10. A significant Analysis of Regression test indicates the relationship is:
    a. likely to be due to skewed means
    b. likely to be due to skewed medians
    c. likely to be due to confounds
    d. unlikely to be due to chance

Progress Check

1. List five measures of dispersion:
    a.
    b.
    c.
    d.
    e.

2. List six criteria for evaluating theories:
    a.
    b.
    c.
    d.
    e.
    f.

3. List six things associated with a linear regression:
    a.
    b.
    c.
    d.
    e.
    f.

4. You wonder if the amount of caffeine has a significant impact on the nervousness of public speakers. Use the following data to complete the summary table:

r= .81
SSx = 14
SSy = 33
N = 9

	SS	df	ms
Regression	___	___	___
Error	___	___	___
TOTAL	___	___	

What is the F for this test:

What is the critical value for F:

Is F significant at .05 alpha: _____

5. As a doctor, you are interested in how similar your patients' patterns of walking and dancing are. You have measured their performance on each task, and now hope to find how related these two variables are. The numbers below represent the number of hours per week spent by the same patients in each activity.

Walking	Dancing
12	2
8	8
9	7
6	4
7	9
4	7
2	12

What is the sum of Walking: _____

What is the SS for Dancing: _____

What is the SSxy: _____

Since you are interested in commonality, which of the following tests should you perform:
      a.multiple regression
      b.regression
      c.correlation
      d.t-test
      e.ANOVA

Perform the comparison you selected in the item above. What was the result of your calculation (select only the appropriate ones):
      $a =$

      $b =$

      $r =$

      $t =$

      $F =$

How many degrees of freedom in this study?

What is the critical value for this statistic?

Is there a significant relationship between these variables at the .05 alpha level?

Calculate the coefficient of determination: _____

6. As a chocolate maker, you are interested in how sugar affects sales. You measure the amount of sugar (tablespoons) in each batch and the amount of chocolate sold (tablespoons).

Sugar	Sales
3	4
1	6
7	11
5	5
14	9

What is the SS for Sugar: _____

What is the SS for Sales: _____

What is the SSxy: _____

Since you are interested in how well one rating acts as a linear predictor of another, which of the following tests should you perform:

    a. t-test
    b. ANOVA
    c. correlation
    d. regression
    e. multiple regression

Perform the comparison you selected in the item above. Select only the appropriate one(s). What was the result of your calculations?

a =

b =

r =

t =

F =

If you put 9 tablespoons of sugar in a batch, how much would you expect to sell?

Answers

Practice Problems

Item 1   Pearson r = .79; F = 6.52

	SS	df	ms
Regression	26.55	1	26.55
Error	16.23	4	4.07
TOTAL	42.83	5	8.57

Item 2   Pearson r = - .73; F = 4.51

	SS	df	ms
Regression	58.74	1	58.74
Error	52.09	4	13.02
TOTAL	110.83	5	22.17

Item 3   F = 8.57, critical value = 4.35, F is significant

	SS	df	ms
Regression	4.50	1	4.50
Error	10.50	20	.53
TOTAL	15.00	21	.71

Item 4   F = 4.06, critical value = 4.49, F is not significant

	SS	df	ms
Regression	3.30	1	3.30
Error	13.00	16	.81
TOTAL	16.30	17	.96

Item 5   F = 11.28, critical value = 4.96, F is significant

	SS	df	ms
Regression	39.22	1	39.22
Error	34.78	10	3.48
TOTAL	74.00	11	

Item 6   F = .16, critical value = 4.84, F is not significant

	SS	df	ms
Regression	.60	1	.60
Error	41.40	11	3.76
TOTAL	42	12	

Simulations

Simulation 1

Alumni: SS = 85.33; sample variance = 17.07; range = 12

SS for Scholarships = 64; SSxy = 60; 01. F = 7.74, critical value = 7.71, F is significant

	SS	df	ms
Regression	42.19	1	42.19
Error	21.81	4	5.45
TOTAL	64.00	5	

Simulation 2

Age: SS = 52; sample variance = 10.40; range = 8

SS for House = 80; SSxy = 62; F = 46.00. Yes, age has a significant impact house price

	SS	df	ms
Regression	73.60	1	73.60
Error	6.40	4	1.60
TOTAL	80.00	5	

Multiple Choice
1. d, 2. a, 3. c, 4. b, 5. c, 6. a, 7. b, 8. d, 9. c, 10. d

Progress Check
1. List five measures of dispersion:
   a. Range
   b. Mean Absolute Deviation
   c. Sum of Squares
   d. Variance
   e. Standard Deviation

2. List six criteria for evaluating theories:
   a. Clear
   b. Useful
   c. Summarizes facts
   d. Small number of assumptions
   e. Internally consistent
   f. Testable hypotheses

3. List six things associated with a linear regression:
   a. Slope
   b. Intercept
   c. Interpolate
   d. Extrapolate
   e. Least squares criterion
   f. Standard error of estimate

4. Analysis of Regression:

	SS	df	ms
Regression	21.78	1	21.78
Error	11.22	7	1.60
TOTAL	33.00	8	

   F = 13.61; critical value = 5.59; F is significant.

5. Correlation
   Sum of Walking = 48; SS for Dancing = 64; SSxy = - 46
   Pearson's r = - .71; df = 5; critical value = .755
   No significant relationship at the .05 alpha level. Coefficient of determination = .51

6. Linear Regression
   SS for Sugar = 100; SS for Sales = 34; SSxy = 36; a = 4.84; b = .36
   If put 9 tablespoons of sugar in a batch, expect to sell = 8.08

Day 8:
t-Tests

Comparing representatives from 2 groups

BRIEFLY

A t-test asks whether two means are significantly different. If the means, as representatives of two samples of the same variable, are equal or close to equal, the assumption is that the differences seen are due to chance. If the means are significantly different, the assumption is that the differences are due to the impact of an independent variable.

When subjects are randomly assigned to groups, the t-test is said to be independent. That is, it tests the impact of an independent variable on a dependent variable. The independent variable is dichotomous (yes/no; treatment/control; high/low) and the dependent variable is continuous. If significant, the independent t-test supports a strong inference of cause-effect.

When subjects are given both conditions (both means are measures of the same subjects at different times), the t-test is said to be dependent or correlated. Because it uses repeated measures, the correlated-t is often replaced by using a regression (where the assumptions of covariance are more clearly stated).

The t-test is a quick and easy way to compare 2 groups. To compare two groups, use:

     Independent t-test

     Correlated t-test

INTRODUCTION

Independent t-test

The independent t-test assumes that one group of subjects has been randomly assigned to 2 groups. Each group contains the same number of subjects, has its own mean and has its own standard deviation.

Conceptually, the t-test is an extension of the z-score. A z score compares the difference between a raw score and the mean of the group to the standard deviation of the group. The result is the number of standard deviations between the score and the group mean.

Similarly, a t-test compares the difference between 2 means to the standard deviation of the pooled variance. That is, one mean pretends to be a raw score and the other mean is the mean of the group. The difference between these means is divided by a standard deviation; it's calculated a little funny but conceptually it's equivalent to the standard deviation used in calculating a z score.

Like a z score, a t-test is evaluated by comparing the calculated value to a standard. In the case of a z score, the standard is the Area Under the Normal Curve. Similarly, a t-test compares its calculated value to a table of critical values. When N is large (infinity, for example), the values in the two tables are identical.

For example, in a one-tailed test at .05 alpha, the critical region would be the top 5% of the distribution. The z-score would be the one where 5% was beyond the z and 45% was between the mean and z (there's another 50% below the mean but the table doesn't include them). The appropriate z-score for the location where there is 5% beyond is 1.65. In the Critical Values of Student's _t_, the critical value at the bottom of the .05 alpha 1-tailed column is 1.65.

Similarly, in a two-tailed test at the .05 alpha, the critical region would be the bottom 2.5% and the top 2.5%. The z-score for the bottom 2.5% is -1.96 and the z-score for the top 2.5% is +1.96. In the Critical Values of Student's _t_ table, the critical value at the bottom of the .05 alpha 2-tailed column is 1.96.

When the t-test has an infinite number of subjects, its critical value is the same as a z-score. At infinity, t-tests could be evaluated by referring to the _Areas Under the Normal Curve_ table. A t-test, however, usually has a small number of subjects. Consequently, the values are modified to adjust for the small sample size.

Significance

The t-test tells us if there is a significant difference between the means. It is as if two armies met and decided that each side would send a representative to battle it out. The representative would not be the best from each side but the average, typical member of their respective groups. Similarly, by comparing the means, we are comparing the representatives of two groups. The entire cast is not involved, only a representative from each side.

We typically do a two-tailed test. That is, we want to know if Group 2 is significantly better than Group 1 AND if it is significantly worse. We want to know both things, so we start at the mean and assume that in order to be significantly different from chance, the t statistic has to be at either of the 2 tails. At .05 alpha (the amount of Type I error we are willing to accept), the critical region is split into two parts, one at each tail. Although the overall alpha level is 5%, there is only 2.5% at each tail.

In one-tailed tests, the entire 5% in a .05 alpha test would be at one end. That is, we would only want to know if Group 2 was significantly better than Group 1; we wouldn't care if it was worse. It doesn't happen very often that our hypotheses are so finely honed that we are interested in only one end of the distribution. We general conduct a 2-tailed test of significance. Consequently, the t statistic might be positive or negative, depending on which mean was put first. There is no theoretical reason why one mean should be placed first in a two-tailed test, so apart from identifying which group did better, the sign of the t-test can be ignored.

Consider the following data:

Group 1	Group 2
8	5
8	3
11	8
14	4
8	7
6	4
4	2

To calculate an independent t-test, we need the mean and Sum of Squares for each variable.
Then we use this formula:

$$t = \frac{\overline{X}_1 - \overline{X}_2}{\sqrt{\dfrac{SS_{x_1} + SS_{x_2}}{n(n-1)}}}$$

Doing the calculations, we find:
SSx (the Sum of Squares of X) = 63.71
SSy (the Sum of Squares of Y) = 27.43
barX (the mean of X) = 8.43
barY (the mean of Y) = 4.71

The n for each variable is 7, so the total number of subjects (N) is 14. The degrees of freedom is N -2, so the df for this study = 12.

Inserting the appropriate information into the appropriate slots in the formula, we calculate that $t = \underline{2.52}$

Comparing this to the critical value of 2.18, which we got by looking up the critical value at the .05 alpha level. We used the <u>Critical Values of Student's *t*</u> table with 12 degrees of freedom.

Since the value we calculated is larger than the book, the t is significant. That is, there is a significant difference between the two groups and they are unlikely to be from the same population.

Correlated t-test

Instead of randomly assigning subjects, some studies reuse people. The advantage is that each person acts as their own control group. Since no one is more like you than you, the control group can't be more like the treatment group. The t-test for repeated measures designs is called a correlated t-test.

The second advantage is that a correlated t-test has more power (is able to use less people to conduct the study). An independent t-test has N-2 degrees of freedom. So if 20 people are randomly assigned to 2 groups, the study has 18 degrees of freedom. In a correlated t-test, if we use all 20 people, the study has 19 degrees of freedom.

The third advantage to correlated designs (also called within-subjects or repeated measures designs) is cost. Reusing people is cheaper. If subjects are paid to participate, they are paid for being in the study, regardless of how many trials it takes. Reusing people is also cheaper in time, materials and logistical effort. Once you have a willing subject, it's hard to let them go.

The primary disadvantage of a correlated t-test is that it is impossible to tell if the effects of receiving one treatment will wear off before receiving the second condition. If people are testing 2 drugs, for example, will the first drug wear off before subjects are given the second drug?

A second problem with the pre- and post-test design often used with correlated t-tests is in its mathematical assumptions. Although the arguments are beyond the scope of this discussion, statisticians differ on the theoretical safety of using difference scores. Some worry that subtracting post-tests from pre-tests may add additional error to the process.

Consequently, a better way of testing correlated conditions is to use a correlation, a linear regression or an analysis of regression. Correlations test for relationship and can be used on ordinal and ratio data. Similarly, linear regression and analysis of regression make predictions and test for goodness of fit without relying on difference scores.

Correlated t-tests are sometimes called repeated-measures of within-subjects designs.

UNDERSTAND

Hypothesis testing

Illustration 1

Hypothesis testing is like venturing out onto a frozen lake. The primary hypothesis is that the lake is frozen but you proceed as if it weren't. You're cautious until you're sure the ice is thick enough to hold you. The H_0 is that the ice is not frozen; this is your null hypothesis (no difference from water). When you have tested the ice (jumping up and down on it or cutting a hole in it to measure the thickness of the ice), you then decide to accept the null hypothesis (no difference from water) or reject that hypothesis and accept the H_1 hypothesis that the lake is frozen and significantly different from water.

Illustration 2

Hypothesis testing is like putting a new roof on your house. You hope that it won't leak (H_1) but you start with the assumption that it will leak (no different from the old roof). Your null hypothesis (H_0) is that it is not better. You then use your hose to test the roof (it's more convenient than waiting for a storm to come). If the roof leaks, you accept the null hypothesis and reject the alternative hypothesis. If the roof doesn't leak, you reject the null hypothesis and accept the alternative (also called experimental) hypothesis.

Illustration 3

Hypothesis testing is like buying a used car. Your null hypothesis is that the car is no different from a new car. Your alternative hypothesis is that it's a lemon. You drive the car, honk the horn and kick the tires. Then you choose between the null hypothesis (no difference from new) or the experimental hypothesis (significant difference from new).

Estimation

We use t-tests to make confidence estimations. When t is significant, we are saying that we are confident that our findings are true 95% of the time (assuming the alpha level is set at .05). Our confidence estimates are interval estimates of a distribution of t scores. A significant t says that its value falls in a restricted part of the distribution (upper 5%, for example).

Illustration 1

Estimation is like getting your car fixed. If you go to a repair shop and they estimate the cost to repair you car is $300, that's a point estimate. An interval estimate would be a range of numbers. If the shop says it will cost between $200-400, that's an interval estimate.

Illustration 2

Estimation is like getting bids for a new roof. If given a single price, you will have received a point estimate. If you are given a range, it is an interval estimate.

Illustration 3

Estimation is like guessing what your golf score will be. A point estimate would be to say you'll score 80. If you guess "between 75 and 80," you are using an interval estimate

REMEMBER
Basic Facts

List and describe nine applications of the General Linear Model for continuous and discrete variables:

Continuous Models compare:

a. Causal modeling	Multiple measures of multiple factors
b. Multivariate analysis	Multiple predictors; multiple criteria
c. Multiple regression	Multiple predictors & single criterion
d. Regression	Single predictor & single criterion
e. Correlation	2 regression lines
f. Frequency distribution	1 variable (predictor or criterion)

Discrete Models compare:

a. T-test 2 means;	1 independent variable
b. One-way ANOVA	2+ means; 1 independent variable
c. Factorial ANOVA	2+ means; 2+ independent variables

Formulas

$$t = \frac{\overline{X_1} - \overline{X_2}}{\sqrt{\dfrac{SS_{x_1} + SS_{x_2}}{n(n-1)}}}$$

Terms:

1-tailed test
2-tailed test
correlated t-test
critical value
df
estimation
hypothesis testing
independent t-test

DO

Step-by-Step

1. Consider this data set

In order to illustrate how to calculate an independent t-test, let's use this data:

Group1	Group2
2	10
3	9
4	12
8	7
5	8
5	13
8	11

2. Calculate the independent t

A t-test seeks to discover if there is a significant difference between two means. This step has six sub-steps:

First, like a z-score, the t-test begins with subtraction. Subtract one mean from another; it doesn't matter which one you start with. The mean of Group1 is 5 and the mean of Group2 is 10. So, the difference between the means = 5 (or -5 if you put the means in the other order).

Second, add the SS for each group together. The SS1 is 32 and the SS2 is 28. So the sum of the two Sum of Squares = 60.

Third, multiply n (the number of people in one group) times n-1. Each group has 7 scores, so 7 times 6 = 42.

Fourth, divide Step 2 by Step 3. That is, 60 divided by 42 = 1.43.

Fifth, square-root Step 4. The square-root of 1.43 = 1.20.

Sixth, divide Step 1 by Step 5. And 5 divided by 1.2 = 4.17. So t = 4.17.

In formal terms, here is how to calculate a t-test

$$t = \frac{\overline{X_1} - \overline{X_2}}{\sqrt{\dfrac{SS_{x_1} + SS_{x_2}}{n(n-1)}}}$$

Although it looks complicated, this formula is quite similar to that of the z-score. The top half of the equation is simply the difference between two means. The bottom half is the standard deviation: the square-root of a pooled variance (combining both groups into one).

3. Find the critical value

The Critical Values of Student's t table is a way to accommodate small sample sizes.

First, calculate N. That is, count the total number of subjects used in the study. Since there were 7 people in each group, combining both groups together equals 14.

Second, enter the table with N-2 degrees of freedom (df). So go down the table to 12 df and across to the .05 alpha level (2-tailed).

The critical value = 2.18. That's the value you have to beat; if your t-test is larger than 2.18, there is a significant difference between the means.

4. Test significance

Since the critical value for our example was 2.18. And the t we calculated was 4.17, t is significant and the differences between the two groups are unlikely to be due to chance.

5. Which group did best

To interpret the t-test, you must know what was being measured (called the dependent variable). If the dependent variable was errors on a test, you probably want the group that has the lowest mean. If the dependent variable was dollars earned, you probably want the group with the largest mean. It is impossible to interpret the results of an independent t-test without knowing what was being measured.

Practice Problems

Item 1

Calculate an independent t-test for the following data:

X_1	X_2
6	12
4	4
2	7
3	10
9	5
6	8
5	3

Mean of group 1 _____

Mean of group 2 _____

Difference between the means_____

SS of group 1 _____

SS of group 2 _____

n(n-1) _____

t = _____

How many degrees of freedom (df) are in this study _____

What is the critical value for t (2 tailed, .05 alpha) _____

Is the t significant _____

Item 2

Calculate an independent t-test for the following data:

X_1	X_2
15	3
11	5
8	4
12	2
7	6

Mean of group 1 _____

Mean of group 2 _____

SS of X_1 _____

SS of X_2 _____

t = _____

Item 3

Calculate an independent t-test for the following data:

X_1	X_2
6	3
7	5
6	3
8	2
4	7
11	4

t = _____

How many degrees of freedom (df) are in this study _____

What is the critical value for t (2 tailed, .05 alpha) _____

Is the t significant? _____

Item 4

Calculate an independent t-test for the following data:

A	B
5	2
11	4
9	2
6	5
2	3

t = _____

Is the t significant (.05 alpha)? _____

Item 5

Which students do significantly better in reading (words per minute):

High Scl	College
7	5
4	3
6	1
6	11
3	8
2	8
1	13
6	9

$t = \underline{\qquad}$

Item 6

Which cars are the safest (number of accidents):

Foreign	Domestic
16	1
14	3
12	9
18	11
12	6
7	1

$t = \underline{\qquad}$

Item 7

Which dogs are the meanest (bites per minute):

Big	Little
1	6
4	3
3	9
8	11
2	16
7	11
5	22
3	7

$t = \underline{\qquad}$

Simulations

Simulation 1

As a coach, you are interested in how well each team has learned to run. You randomly assigned your players to two teams. Team A trained the old fashioned way. Team B is using computer-assisted training. The numbers in the columns below are the minutes needed to run around the track once.

Team A	Team B
16	1
7	3
2	4
4	4
4	2
8	4
8	3

What is the median for Team A:_____

What is the mean for Team A: _____

What is the SS for Team A: _____

What is the mode for Team B: _____

What is the SS for Team B: _____

Since you are interested in which group did best, which of the following tests should you perform:

    a.t-test
    b.ANOVA
    c.correlation
    d.regression
    e.multiple regression

Perform the comparison you selected in the item above. Select only the appropriate one(s). What was the result of your calculation?

    $a =$

    $b =$

    $r =$

    $t =$

    $F =$

How many degrees of freedom are in this study?

Which training program will you suggest to use in the future?

Simulation 2

2. As a merchant, you are interested in how well your coupon works. You randomly assigned your customers to two groups. Half receive 10% off their next purchase, and half receive a free movie ticket. You measure the number of dollars of profit per day. Which is the better promotion?

10% Off	Free Movie
2	11
1	24
3	7
2	9
4	22
6	12
3	17
3	14

What is the mean for 10% Off: _____

What is the SS for 10% Off: _____

What is the mode for Movie: _____

What is the SS for Movie: _____

Since you are interested in which group did best, which of the following tests should you perform:

       a.t-test
       b.ANOVA
       c.correlation
       d.regression
       e.multiple regression

Perform the comparison you selected in the item above. Select only the appropriate one(s). What was the result of your calculation?

       $a =$

       $b =$

       $r =$

       $t =$

       $F =$

Which promotion is best for sales?

SUMMARY

There are two kinds of t-tests: independent and correlated.

Correlated t-tests use paired observations on one group of subjects. This is called a within-subjects design.

Independent t-tests use subjects which have been randomly assigned to two groups.

The degrees of freedom for a correlated t-test equals n-1.

The degrees of freedom for an independent t-test equals N-2.

T-tests are not used with more than two groups because of the likelihood of Type I error.

T-tests measure the differences between means.

T-tests are like z-scores.

The independent t-test pools the variance of the subgroups.

1. A null hypothesis says that two means are:
   a. significantly different
   b. slight different
   c. not significantly different
   d. you have hit the null on the head

2. A t-test compares:
   a. two medians
   b. two modes
   c. two means
   d. two standard deviations

3. A t-test is calculated like which of the following:
   a. point estimation
   b. sum of squares
   c. degrees of freedom
   d. z-score

4. Rejecting the null when you should have accepted it is a:
   a. Type I error
   b. Type II error
   c. Type III error
   d. Typo error

5. Compared to an alpha level of .20, an alpha level of .05 is <u>less likely</u> to have:
   a. Type I error
   b. Type II error
   c. Type III error
   d. Type IV error

6. When the null hypothesis is true, the expected value for an independent measures t statistic is:
   a. 0
   b. +1.0
   c. -1.0
   d. +1 or -1

7. An independent t-test has degrees of freedom equal to:
   a. N
   b. N-1
   c. N-2
   d. can't tell

8. How many dependent variables does a t-test have:
   a. 1
   b. 2
   c. 3
   d. N

9. Which is an advantage of randomly assigning subjects, compared to repeated measures designs:
   a. cheaper to run
   b. more power
   c. less chance of confounds
   d. all of the above

10. What's the critical value for an independent t-test with 42 subjects:
   a. .69
   b. 1.86
   c. 1.96
   d. 2.02

Progress Check

1. List and describe nine applications of the General Linear Model for continuous and discrete variables:

Continuous Models compare:

a.

b.

c.

d.

e.

f.

Discrete Models compare:

a.

b.

c.

2. As a clinician, you are interested in the relationship between the amount of sleep your patients get and their sense of hope. You measure all of your patients on these two variables:

Sleep	Hope
17	21
12	12
10	9
6	7
4	3
1	2

What is the sum of Sleep: _____

What is the SS of Sleep: _____

The variance of Sleep is: _____

What is the mean of Hope: _____

What is the SS of Hope: _____

Since you are interested in commonality, which of the following tests should you perform:
     a.t-test
     b.ANOVA
     c.correlation
     d.regression
     e.multiple regression

Perform the comparison you selected in the item above. Select only the appropriate one(s). What was the result of your calculation?

     $a =$

     $b =$

     $r =$

     $t =$

     $F =$

Calculate the coefficient of determination:

What percentage of variance is unaccounted for:

3. As a director of sales, you are interested in how well your sales did over the year. In particular, you are curious whether last year's revenue is a good predictor of this year's. The numbers below are the number of dollars (in millions) for some of the months.

Last Year	This Year
1	16
3	11
5	8
7	5
7	4
7	2
12	- 7

What is the SS for Last Year: _____

What is the variance of Last Yr:_____

What is the range of Last Year:_____

What is the SS for This Year: _____

What is the SSxy: _____

Since you are interested in how well one test acts as a linear predictor of another, which of the following tests should you perform:
   a.   t-test
   b.   ANOVA
   c.   correlation
   d.   regression
   e.   multiple regression

Perform the comparison you selected in the item above. Select only the appropriate one(s). What was the result of your calculations?

a =

b =

r =

t =

F =

If a client's last year revenue was 14 million, what would you predict this year's revenue to be?

4. As a sculling coach, you are interested in how well your athletes have learned to row. You randomly assigned your players to two group. Group A trained by paddling logs around a lake. Group B used ergs in the gym. The numbers in the columns below are the minutes needed to row 5000 meters.

Logs	Ergs
22	4
14	3
16	10
8	4
15	4

What is the median for Logs: _____

What is the mean for Logs: _____

What is the SS for Logs: _____

What is the mode for Ergs: _____

What is the SS for Ergs: _____

Since you are interested in which group did best, which of the following tests should you perform:

      a.t-test
      b.ANOVA
      c.correlation
      d.regression
      e.multiple regression

Perform the comparison you selected on the item above. Select only the appropriate one(s). What was the result of your calculation?

      $a =$

      $b =$

      $r =$

      $t =$

      $F =$

Which training program will you suggest to use in the future?

5. You want to know if the amount of advertising has a significant impact on the sale of your widgets. Use the following data to complete the summary table:

$r = -.62$
$SS_x = 21$
$SS_y = 13$
$N = 18$

	SS	df	ms
Regress	5.00	1	5.00
Error	8.00	16	0.50
TOTAL	13.00	17	

What is the criterion in this study: sales of widgets

What is the F for this test: 10.00

What is the critical value for F: 4.49

Is F significant at .05 alpha: yes

Answers

Practice Problems

Item 1

Mean of group 1 = 5

Mean of group 2 = 7

Difference between the means = 2

SS of group 1   32

SS of group 2   64

$n(n-1)$ 42

$t =$ 1.32

How many degrees of freedom (df) are in this study: 14-2 = 12

What is the critical value for t (2 tailed, .05 alpha):   2.18

Is the t significant: No

Item 2   t = 4.13

Mean of group 1   10.60

Mean of group 2   4

SS of X_1 41.20

SS of X_2 10

Item 3   t = 2.48

Degrees of freedom (df) are in this study: 10

What is the critical value for t (2 tailed, .05 alpha): 2.23

Is the t significant: Yes

Item 4   t = 2.03.

Is the t significant (.05 alpha): No

Item 5   t = 1.77

Item 6   t = 3.45

Item 7   t = 2.86

Simulations

Simulation 1   t-test

Team A: median = 7; mean = 7.00; SS = 12.60

Team B: mode = 4; SS = 8

t = 2.23   Degrees of freedom (df) = 12

Which training program will you suggest to use in the future?

Team B is significantly better (use computer-assisted training)

Simulation 2   t-test

10% Off: mean = 3; SS = 16

Movie: mode = none; SS = 258

t = 5.20

Which promotion: free movie earned significantly more dollars

Multiple Choice

1. c, 2. c, 3. d, 4. a, 5. a, 6. a, 7. c, 8. a, 9. c, 10. d

Progress Check

1. Nine applications of the General Linear Model for continuous and discrete variables:

Continuous Models compare:

a. Causal modeling	Multiple measures of multiple factors
b. Multivariate analysis	Multiple predictors; multiple criteria
c. Multiple regression	Multiple predictors & single criterion
d. Regression	Single predictor & single criterion
e. Correlation	2 regression lines
f. Frequency distribution	1 variable (predictor or criterion)

Discrete Models compare:

a. T-test 2 means;	1 independent variable
b. One-way ANOVA	2+ means; 1 independent variable
c. Factorial ANOVA	2+ means; 2+ independent variables

2. Correlation

Sleep: sum = 50; SS =169.33; population variance = 28.22

Hope: mean = 9; SS = 242

$r = .97$ Coefficient of determination = .95   Percent of unaccounted variance = .05

3. Regression

Last Year: SS = 74; sample variance = 12.33; range = 11

This Year: SS = 317.71

$SS_{xy} = -152$ $a = 17.90$ $b = -2.05$

If X = 14 million, predict this year's revenue to be: - 10.86 million

4. t-test

Logs: median =15; mean = 15; SS = 100

Ergs: mode = 4; SS = 32

$t = 3.89$ Critical value = 2.31

Which training program = ergs is significantly better

Item 4   Analysis of Regression

	SS	df	ms
Between	5	1	5
Within	8	16	.50
TOTAL	13	17	

Criterion in this study = Sales

$F = 9.99$ Critical value = 4.49   Yes, it is significant at .05 alpha level

Day 9:
1-Way ANOVA
Comparing 3+ groups

BRIEFLY

When more than 2 groups are to be compared, multiple t-tests are not conducted because of the increased likelihood of Type I error. Instead, before subgroup comparisons are made, the variance of the entire design is analyzed. This pre-analysis is called an Analysis of Variance (ANOVA for short). Using the F-test (like an Analysis of Regression), an ANOVA makes a ratio of variance between the subgroups (due to the manipulation of the experimenter) to variance within the subgroups (due to chance).

Comparing 3 or more groups requires a pre-analysis of the data. To compare a group to a standard, use:

Within-Subjects Designs
1-Way ANOVA

INTRODUCTION
Within-subjects

Sometimes we want to take repeated measures of the same people over time. These specialized studies are called *within-subjects* or *repeated measures* designs. Conceptually, they are extensions of the correlated t-test; the means are compared over time.

Like correlated t-tests, the advantages are that subjects act as their own controls, eliminating the difficulty of matching subjects on similar backgrounds, skills, experience, etc. Also, within-subject designs have more power (require less people to find a significant difference) and consequently are cheaper to run (assuming you're paying your subjects).

They also suffer from the same disadvantages. There is no way of knowing if the effects of trial 1 wear off before the subjects get trial 2. The more trials in a study the larger the potential problem. In a multi-trial study, the treatment conditions could be impossibly confounded.

A more detailed investigation of within-subject designs is beyond the scope of this discussion. For now, realize that it is possible, and sometimes desirable, to construct designs with repeated measures on the same subjects. But it is not a straightforward proposition and requires more than an elementary understanding of statistics.

One-Way ANOVA

It is called one-way because there is one independent variable is this design. It is called an ANOVA because that's an acrostic for ANalysis Of VAriance. An 1-way analysis of variance is a pre-test to prevent Type I error.

Although we try to control Type I error by setting our alpha level at a reasonable level of error (typically 5%) for one test, when we do several tests, we run into increased risk of seeing relationships that don't exist. One t-test has a 5/100 chance of having Type I error. But multiple t-tests on the same data set destroy the careful controls we set in place.

We can use a t-test to compare the means of two groups. But to compare 3, 4 or more groups, we'd have to do too many t-tests; so many that we'd risk finding a significant t-test when none existed. If there were 4 groups (A, B, C and D, we'll call them), to compare each condition to another you'd have to make the following t-tests:

AB
AC
AD
BC
BD
CD

The chances are good that we'll find one of those tests look significant but will not be. What we need is a pre-analysis of data to test the overall design and then go back, if the overall variance is significant, and conduct the t-tests.

Theory of F

The premise of an ANOVA is to compare the amount of variance between the groups to the variance within the groups.

The variance within any given group is assumed to be due to chance (one subject had a good day, one was naturally better, one ran into a wall on the way out the door, etc.). There is no pattern to such variation; it is all determined by chance.

If no experimental conditions are imposed, it is assumed that the variance between the groups would also be due to chance. Since subjects are randomly assigned to the groups, there is no reason other than chance that one group would perform better than another.

After the independent variable is manipulated, the differences between the groups are due to chance and the independent variable. By dividing the between group variance by the within variance, the chance parts should cancel each other out. The result should be a measure of the impact the independent variable had on the dependent variable. At least that's the theory behind the F test.

Significance

Yes, this is the same F test we used doing an Analysis of Regression. And it has the same summary table:

	SS	df	ms
Between	___	___	___
Within	___	___	___
Total	___	___	___

Notice that the titles have changed. We now talk about Between Sum of Squares, not Regression SS. The F test is the ratio of between-group variance (called between mean squares or mean squares$_{between}$) to within-group variance (called within mean squares or mean squares$_{within}$).

After you calculate the F, you compare it to the critical value in a table of Critical Values of F. There are several pages of critical values to choose from because the shape of F distribution changes as the number of subjects in the study decreases. To find the right critical value, go across the degrees of freedom for regression (df$_{between}$) and down the df$_{within}$.

Remember, we're using the same procedures and rationale we talked about on Day 7. Analysis of Regression and Analysis of Variance use the same F test. Once you find F, the process of determining significance and the interpretation of the results is the same. You already know how to do this.

As you'll recall, we set the alpha level at .05 to make sure that we make the right decision. We can't know for sure that we're making the right decision; our knowledge is not complete. We're guessing but we're basing our predictions on data. We start with the assumption that all of the relationships we think we see are due to chance. And we keep that assumption until we are 95% sure that a relationship exists. We also describe the relationship as "significant;" we ignore mild, casual, weak and temporary relationships. We're only interested in relationships that are so strong we must call them significant.

Example

Group 1	Group 2	Group 3
6	6	13
4	1	10
5	3	6
8	2	7
2	8	9

In order to calculate an Analysis of Variance for this data, we fill in the blanks for the Analysis of Variance's summary table:

	SS	df	ms
Between	____	____	____
Within	____	____	____
Total	____	____	

Let's start with the degrees of freedom. Like ANOR, an ANOVA has degrees of freedom. The $df_{between} = k-1$, where k is the number of columns (kolumns) or groups (kroups?). Three columns minus 1 = 2.

We know that df_{within} is equal to N-k (the number of people minus the number of columns); so 15-3 = 12. The total degrees of freedom is equal to N-1; 15-1 = 14.

With this in mind, let's update the summary table with what we know:

	SS	df	ms
Between	____	2	____
Within	____	12	____
Total	____	14	

Calculating SS

To calculate the Sum of Squares for this study, we begin with collecting some summary information. Find the sum of each group, the n of each group, X-squares of each group (square each score and add them up), and the SS for each group. Then, add each column across and put it in a Totals column. Like this:

ANOVA	X1	X2	X3	TOTALS
	6	6	13	
	4	1	10	
	5	3	6	
	8	2	7	
	2	8	9	
ΣX	25	20	45	90
ΣX^2	145	114	435	694
N	5	5	5	15
SS	20.00	34.00	30.00	84

SS_{within}

The SS for each group (20, 34 and 30, respectively) is the amount of dispersion WITHIN that group. So the sum of those SS, gives us Within SS (84 in this case). That is, SS_{within} is the $SS_{x1} + SS_{x2} + SS_{x3}$....

$SS_{between}$

The $SS_{between}$ is a bit more challenging to calculate. Here is the formula for it:

$$SS_b = \sum\left(\frac{(\sum AB)^2}{n}\right) - \frac{G^2}{N}$$

Impressive, huh? Let me explain so it's not so scary. Take the sum of each group (that's the AB part of the equation):

$25^2 + 20^2 + 45^2$

Divide it by the number of scores in each group (n). NOT the number of columns, the number of scores:

$$(25^2 + 20^2 + 45^2)/5$$

Now square the total of all the raw scores (G for grand total of X) and divide it by N (the number of people in the study). And subtract this from the subgroup numbers:

$$\frac{25^2 + 20^2 + 45^2}{5} \quad \text{minus} \quad \frac{90^2}{15}$$

This is going to give us:

$$\frac{625 + 400 + 2025}{5} \quad \text{minus} \quad \frac{8100}{15}$$

And this gives us: 3050 / 5 minus 540. That is:

610 minus 540 = 70. The $SS_{between}$ equals 70.

Getting to F

SS_{total}

To find the SS_{total} we use the totals column and plug those numbers into the regular Sum of Squares formula. That is, we ignore which group the scores come from and treat them as if they were all in one group. So, 694 minus (90^2) / 15. Notice that the last part of the formula has already been calculated when we did the $SS_{between}$. What we come to, then, is 694 minus 540, which equals 154. The SS_{total} = 154.

Updating the summary table gives us:

	SS	df	ms
Between	70.00	2	35.00
Within	84.00	12	7.00
Total	154.00	14	11.00

Finding F

F is mean squares between ($ms_{between}$) divided by mean squares within (ms_{within}). In this case, F = 35 / 7 = 5.00

To test the significance of this F, we look up the critical value for the test at 2 and 12 degrees of freedom. Using the F Table, we find that the critical value at .05 alpha is 3.88. Since the value we calculated is larger than the one in the book, F is significant.

Interpretation

Since the F is significant, what do we do now?

Now, all of those t-tests we couldn't do because we were afraid of Type I error are available for our calculating pleasure. So we do t-tests between:

AB

AC

BC

We might find that there is a significant difference between each group, such as this:

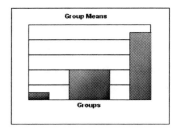

In this example the lowest group is significantly different from the middle group. And the middle group is significantly different from the highest group. And, just to be complete, the lowest group is significantly different from the highest group. That is, A is significantly different from B, and B is significantly different from C. The t-tests of AB, AC and BC would all be significant.

We might find that there is a not a significant difference between two of the groups but that there is a significant difference between them and the third group:

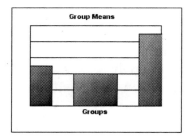 or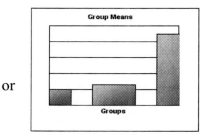

In this case A and B are not significantly different from each other but both significantly differ from C. The t-test of AB would not be significant. But the t-tests of AC and BC would be significant.

Also keep in mind that which group did best depends on whether the numbers are of money (we want the higher means) or errors (we want the lower means). Data requires interpretation because we use a variety of outcome measures. Sometimes we measure things we want more of (e.g., self-esteem, kindness, generosity, health). Sometimes we measure things we don't want to have (errors, fear, anxiety, depression). Numbers won't tell us by themselves. We have to translate them into meaningful interpretations.

If the F had not been significant, there would not be anything left to do. We would have stopped with the calculating of F and concluded that the differences we see are due to chance.

UNDERSTAND

Degrees of freedom

Illustration 1: Degrees of freedom is like having a ball and chain. The mean is the ball and the chain allows you to go only a limited number of steps (degrees of freedom). The chain length can vary from N to N-1 or N-2, etc.

Illustration 2: Degrees of freedom is like being in jail. Scores are free to anywhere...as long as they stay within their degrees of freedom.

Illustration 3: Degrees of freedom is like being on a leash. The mean restricts only a bit (usually 1 or 2 degrees). All the other scores are free to vary.

REMEMBER

Basic Facts

A 1-way ANOVA compares 3+ means on one independent variable.

Formulas:

There are two methods of calculating Sum of Squares: deviation (which is done to

$$SS_b = \sum \left(\frac{(\sum AB)^2}{n} \right) - \frac{G^2}{N}$$

$$SS_t = \sum X^2_1 + \sum X^2_2 + \sum X^2_3 + ... - \frac{(\sum X)^2}{N}$$

Mean Square = SS/df

Terms:

1-Way ANOVA
ANOVA
confound
controls
F
mean squares
$ms_{between}$
ms_{within}
repeated measures design
SS within
$SS_{between}$
SS_{total}
within-subjects design

DO

Step By Step

1. Summarize the subgroups

First, find the n (number of score) for each group.

Second, find the sum for each group.

Third, square each number in the first group and sum them. Then square and sum the scores in each of the other groups.

Fourth, find the Sum of Squares (SS) for each group.

Fifth, find the totals for n, sum, squares, and SS. So, N = 20 (the sum of each group's n's). The SumX = 117 and so on.

Updating our example, it would look like this:

	Group1	Group2	Group3	Group4	Totals
	1	6	12	5	
	2	4	7	2	
	4	9	15	6	
	3	5	9	3	
	2	11	7	4	
n	5	5	5	5	20
Sum	12	35	50	20	117
Squares	34	279	548	90	951
SS	5.20	34	48	10	97.20

2. Find SS$_{within}$

First, start by creating a summary table.

Second, write in the SSwithin. In the process of summarizing the groups, the SSwithin was calculated. It was 97.2. That is, SSwithin (within the experiment) is the sum of the SS that is in (within) each group.

Updating our summary table, it now looks like this:

	SS	df	ms
Between	___	___	___
Within	97.20	___	___
Total	___	___	___

3. Find SS$_{between}$

The formula for SSbetween looks more difficult than it is. Here's the formula:

$$SS_b = \left[\frac{(\sum x_1)^2 + (\sum x_2)^2 + (\sum x_3)^2 + \ldots}{n}\right] - \frac{(\sum x)^2}{N}$$

Here's what to do. First, start with the sum of each group (12, 35, 50, 20). Square each of them and add them together:

$$12^2 + 35^2 + 50^2 + 20^2$$

So we get: 144 + 1225 + 2500 + 400 = 4269.

Second, divide Step1 by the n (the number of subjects in each group. NOTE: It is not the number of groups but the number of scores in each group. This gives us: 4269 divided by 5 = 853.8.

Third, take the Sum of X in the totals column (117) and square it, which equals 13689.

Fourth, divide Step3 by the N in the totals column (20). That is, 13689 divided by 20 = 684.5

Fifth, subtract Step4 from Step2. So, 853.8 - 684.5 = 169.35. This is the SSbetween.

Updating our summary table, it now looks like this:

	SS	df	ms
Between	169.35	___	___
Within	97.20	___	___
Total	___	___	___

4. Find SS$_{total}$

The formula for SStotal is the same as any basic SS. We use the information in the totals column and apply this formula:

$$SS = \sum X^2 - \frac{(\sum X)^2}{N}$$

First, note that the sum of X-squares = 951.

Second, take the sum of X's (117) and square it. This equals 13689.

Third, divide Step2 by 20 (big N), which equals 684.5

Fourth, subtract Step3 from Step1. That is, 951 - 684.5 = 266.55. This is the SStotal.

Updating our summary table, it now looks like this:

	SS	df	ms
Between	169.35	___	___
Within	97.20	___	___
Total	266.55	___	___

To check the calculations, simply add SSbetween to SSwithin and see if they equal SStotal. It does, so we calculated everything correctly.

5. Find the degrees of freedom

First, enter the degrees of freedom (df) for Between, which is k-1 (columns minus one). Since our example has 4 columns, the df for Between = 3.

Second, enter the df for Within, which is N-k (number of people minus the number of columns). In our example, N = 20, so df$_{within}$ = 16.

Third, enter the df for Total, which is N-1 (number of people minus one). So, df$_{total}$ = 19.

	SS	df	ms
Between	169.35	3	___
Within	97.20	16	___
Total	266.55	19	___

6. Find F

First, calculate the appropriate mean squares. Since mean squares is another name for variance (and SS divided by df equals variance), divide each SS by its respective df. Updating the table, we now have:

	SS	df	ms
Between	169.35	3	56.45
Within	97.20	16	6.08
Total	266.55	19	14.03

Second, divide the mean squares of Between by the mean squares of Within. That is, 56.45 divided by 6.08 = 9.29. This ratio is called the F test, so F = 9.29.

7. Find the critical value

The <u>Critical Values of the F Distribution</u> table is actually a series of distributions. To enter the table, go across to the row whose number matches the degrees of freedom for Between ($df_{between}$). And go down the df_{within}.

In our example, go across to 3 and down to 16. The critical value (the value you have to beat) = 3.24 (at the .05 alpha level).

8. Decide what to do next

If F is not significant, there is nothing else to do. The differences between the groups is due to chance.

If F is significant, then t-tests are done: one between each pair of combinations (AB, AC, AD, BC, BD and CD).

To test for significance, the calculated value is compared to the F table. If the value you calculated is bigger than the value in the book, F is significant. In our example, we calculated F to be 9.29, which is bigger than the critical value of 3.24 we found at 3 and 16 degrees of freedom. So, F is significant and the t-tests are authorized.

Practice Problems

Item 1

Which grade has the most car accidents:

10th	11th	12th
2	13	4
9	17	8
3	14	2
1	9	1
7	1	4

	SS	df	ms
Between	____	____	____
Within	____	____	____
Total	____	____	____

Item 2
Which color of house is lived in the longest (in years)?

Blue	Green	Peach
8	11	4
7	9	8
3	7	9
1	18	2
9	12	4

	SS	df	ms
Between	____	___	____
Within	____	___	____
Total	____	___	____

Item 3
Which is the best pizza (most pepperoni):

PHome	PMad	Mo-Pizza
12	3	5
2	3	5
8	0	4
1	1	5
2	2	5
9	4	5

	SS	df	ms
Between	____	___	____
Within	____	___	____
Total	____	___	____

Item 4

Which toothpaste lasts more days?

Brand-X	ToothE	HappyUp
3	1	9
6	7	9
5	3	8
7	6	8
2	5	9

	SS	df	ms
Between	____	____	____
Within	____	____	____
Total	____	____	____

Item 5

Which shoe has the most stripes?

Brand-X	MyTooth	HappyUp
8	3	2
2	4	7
8	4	1
8	7	2
2	2	2

	SS	df	ms
Between	____	____	____
Within	____	____	____
Total	____	____	____

$F =$

What is the critical value for $F =$

Is F significant?

Simulations
Simulation 1

You work for a bottled-water company that is concerned with the quality of its flavored product. The numbers below are the number of complaints received in a 1-hour period.

Lemon	Orange	Lime	Root Beer
12	1	18	1
13	4	9	2
11	5	12	1
14	3	7	9

Since you are interested in which flavor did best, which of the following tests should you perform:

    a.t-test

    b.ANOVA

    c.correlation

    d.regression

    e.multiple regression

What is the independent variable in this design:

What is the dependent variable in this design:

Based on the above data, complete the following summary table:

	SS	df	ms
Between	____	___	____
Within	____	___	____
Total	____	___	____

F =

What is the critical value for F =

Is F significant?

What should be done next?

Simulation 2

Several samples are taken of each flavor of cough syrup. Which flavor tastes the best (number of complaints):

Peach	Strawberry	Melon	Lemonade
4	2	4	1
3	1	8	8
7	7	7	3
5	6	3	9
2	2	7	12

Since you are interested in which flavor did best, which of the following tests should you perform:

a.t-test
b.ANOVA
c.correlation
d.regression
e.multiple regression

What is the independent variable in this design:

What is the dependent variable in this design:

Based on the above data, complete the following summary table:

	SS	df	ms
Between	___	__	___
Within	___	__	___
Total	___	__	___

F =

What is the critical value for F =

Is F significant?

What should be done next?

SUMMARY

Although there are multiple groups, they vary on one independent variable.

The dependent variable is what the numbers measure. The numbers are dependent on the performance of the subjects.

The independent variable is what the experimenter manipulates. It is independent of the subject's performance.

The F is a ratio of the variance between the groups to the variation within the groups.

The F test assumes that the variation within a group is due to ability and chance.

The F test assumes that the variation between groups is due to ability, chance, and manipulation of the independent variable.

The F test assumes that variance due to ability and chance (between and with subjects) will cancel each other out, so that what remains is a measurement of the effect of the independent variable on the dependent variable.

Mean Squares is a variance term.

Mean Squares equals Sum of Squares divided by its appropriate degrees of freedom (SS/df)

Between-Subjects degrees of freedom equals the number of groups minus 1 (k-1).

Within-Subjects degrees of freedom equals the total number of people minus the number of groups (N-k).

Total degrees of freedom equals the total number of people minus 1 (N-1).

ANOVA stands for ANalysis Of VAriance.

1. When more than two levels of one independent variable are to be compared, which of the following should be used:
   a. regression
   b. one-way ANOVA
   c. t-test
   d. factorial ANOVA
   e. all of the above

2. The advantage of a repeated-measures design is that it reduces the contribution of error variability due to:
   a. mean of D
   b. degrees of freedom
   c. the effect of the treatment
   d. individual differences
   e. none of the above

3. The F test is best understood as a:
   a. ratio
   b. confound
   c. standard deviation
   d. decision error
   e. prediction error

4. The total degrees of freedom for a 1-way ANOVA equal:
   a. k
   b. k-1
   c. N
   d. N-1
   e. N-k

5. The df for $SS_{between}$ equals:
   a. k
   b. k-1
   c. N
   d. N-1
   e. N-k

6. How many independent variables are in a 1-way ANOVA:
   a. 1
   b. 3
   c. 6
   d. 9
   e. varies

7. How many dependent variables are in a 1-way ANOVA:
   a. 1
   b. 3
   c. 6
   d. 9
   e. varies

8. For an F-test, which of the following goes on top (numerator):
   a. ms_{total}
   b. ms_{within}
   c. $ms_{between}$
   d. $SS_{between}$
   e. SS_{within}

9. Which of the following should be used to compare three or more groups:
   a. t
   b. F
   c. r
   d. r^2
   e. z

10. Which of the following is an estimate of error:
    a. SS_{total}
    b. $SS_{between}$
    c. SS_{within}
    d. t
    e. r

Progress Check

1. List three measures of central tendency:

   a.

   b.

   c.

2. List five measures of dispersion:

   a.

   b.

   c.

   d.

   e.

3. List six criteria for evaluating theories:

   a.

   b.

   c.

   d.

   e.

   f.

4. List four levels of measurement:

   a.

   b.

   c.

   d.

5. As a director of personnel, you are curious whether peer ratings are a good predictor of supervisor ratings. The numbers below are a sample of the company's data.

Supervisor	Peer
6	2
3	2
2	4
4	1
5	2
5	2
6	3
6	8
6	7

What is the SS for Peer Ratings: _____

What is the variance of Peer Ratings: _____

What is the range of Supervisor Ratings: _____

What is the SS for Supervisor Ratings: _____

What is the SSxy: _____

Since you are interested in how well one rating acts as a linear predictor of another, which of the following tests should you perform:

a. t-test

b. ANOVA

c. correlation

d. regression

e. multiple regression

Perform the comparison you selected in the item above. Select only the appropriate one(s). What was the result of your calculations?

a =

b =

r =

t =

F =

If a peer rating was 11, what would you predict their supervisor rating to be?

6. As a teacher, you are interested in how reading and spelling are related. You have measured all of your students in each area and do not wish to generalize beyond your class. You hope to find how interrelated these two variables are. Below are the number correct on each test.

Reading	Spelling
19	3
11	14
13	15
14	16
6	13
8	12
8	5

What is the sum of Reading: _____

What is the SS of Spelling: _____

The variance of Reading is: _____

What is the mean of Spelling: _____

What is the SS of Reading: _____

Since you are interested in commonality, which of the following tests should you perform:
    a.t-test
    b.ANOVA
    c.correlation
    d.regression
    e.multiple regression

Perform the comparison you selected in the item above. Select only the appropriate one(s). What was the result of your calculations?

    a =

    b =

    r =

    t =

    F =

What is the critical value at .05 alpha:

Is this test significant:

7. As a manager, you randomly assigned your staff to two locations. The numbers in the columns below are dollars. Which is the significantly better location?

New York	LA
8	11
2.4	9
6	3
4	9
4	9
5	12
3	14
9.2	7.2

What is the median for New York: _____

What is the mean for New York: _____

What is the SS for New York: _____

What is the mode for Los Angeles: _____

What is the SS for Los Angeles: _____

Since you are interested in which group did best, which of the following tests should you perform:

      a.t-test
      b.ANOVA
      c.correlation
      d.regression
      e.multiple regression

Perform the comparison you selected in the item above. Select only the appropriate one(s). What was the result of your calculations?

      a =

      b =

      r =

      t =

      F =

Which location is best for sales?

8. You work for a cookie company which is testing a new product. Each group is composed of independent subjects, randomly assigned to a treatment by you. The four groups differ in the amount of sugar in the cookies. The numbers below are the number of cookies eaten in a 2 hour period.

Sugarless	Lo Sugar	Medium	High
7	6	8	16
3	6	8	12
5	4	5	11
3	3	5	15
4	3	3	11

Since you are interested in which cookie did best, which of the following tests should you perform:
   a. t-test
   b. ANOVA
   c. correlation
   d. regression
   e. multiple regression

How many independent variables are in this design: _____

How many dependent variables are in this design: _____

Based on above data, complete the following summary table:

	SS	df	ms
Between	____	____	____
Within	____	____	____
TOTAL	____	____	

Calculate the F for this test:

What is the critical value:

Is the F significant?

What should be done next?

You wonder if the amount of salt has an significant impact on the rating of your ice cream's quality. Use the following data to complete the summary table:

r= .44
SSx = 122
SSy = 40
N = 14

	SS	df	ms
Regression	____	____	____
Error	____	____	____
TOTAL	____	____	

What is the predictor in this study:

What is the F for this test: _____

What is the critical value for F: _____

Is F significant at .05 alpha: _____

Answers

Practice Problems

Item 1   F = 3.95

	SS	df	ms
Between	150.53	2	75.27
Within	228.80	12	19.07
Total	379.33	14	27.10

Item 2   F = 4.60

	SS	df	ms
Between	116.13	2	58.07
Within	151.60	12	12.63
Total	267.73	14	19.12

Item 3   F = 2.57

	SS	df	ms
Between	40.11	2	20.06
Within	116.99	15	7.80
Total	157.11	17	9.24

Item 4   F = 8.10

	SS	df	ms
Between	56.13	2	28.07
Within	41.60	12	3.47
Total	97.73	14	6.98

Item 5   F = 1.48

	SS	df	ms
Between	19.73	2	9.87
Within	80.00	12	6.67
Total	99.73	14	7.12

Critical value = 3.88. F is not significant

Simulations

Sim 1   F = 9.67

	SS	df	ms
Between	308.25	3	102.75
Within	127.50	12	10.63
Total	435.75	15	29.05

This is an ANOVA. Critical value = 3.49. F is significant at
    the .05 alpha level. T-tests should be done next.
Independent variable = flavor.
Dependent variable = # of compliments

Simulation 2   ANOVA

Sim 2   F = 1.07

	SS	df	ms
Between	28.95	3	9.65
Within	144	16	9
Total	172.95	19	9.10

This is an ANOVA. Critical value = 3.24. F is not significant at
    the .05 alpha level. Do nothing more.
Independent variable = flavor.
Dependent variable = # of compliments

Multiple Choice

1. b, 2. d, 3. a, 4. d, 5. b, 6. a, 7. a, 8. c, 9. b, 10. c

Progress Check
1. mean, median and mode
2. range, MAD, Sum of Squares, variance and standard deviation
3. clear, useful, summarize facts, small number of assumption, internally consistent, and testable hypotheses
4. nominal, ordinal, interval and ratio
5. Regression
    Peer (X) predicts Supervisor (Y)
    What is the SS for Peer Ratings: 48.22
    What is the variance of Peer Ratings: 6.03
    What is the range of Supervisor Ratings: 4
    What is the SS for Supervisor Ratings: 17.56
    What is the SSxy: 9.89
    a = 4.07   b = .205   If a peer rating was 11, predict supervisor rating to be: 6.33

6. Correlation
    Sum of Reading: 79
    Sum of Squares of Reading: 119.43
    Variance of Reading: 17.06 (population)
    Mean of Spelling: 11.14
    SS of Spelling   154.86
    r = - .27   Critical value = .755   Not significant at .05 alpha (5 df)

7. t-test
    What is the median for New York: 4.50
    What is the mean for New York: 5.20
    What is the SS for New York: 40.08
    What is the mode for Los Angeles: 9
    What is the SS for Los Angeles: 76.64
    t = 2.82   Critical value = 2.15 at 14 df; sig. at .05 alpha level
    Which location is best for sales?   Los Angeles

8. 1-Way ANOVA
    1 independent variable; 1 dependent variable

	SS	df	ms
Between	254.60	3	84.87
Within	61.20	16	3.83
TOTAL	315.80	19	

    F = 22.19   Critical value = 3.24   Yes, significant. Do t-tests

9. Analysis of regression

	SS	df	ms
Between	7.74	1	7.74
Within	32.26	12	2.69
TOTAL	40	13	

    What is the predictor in this study: Salt
    What is the F for this test: 2.88
    What is the critical value for F: 4.75
    Is F significant at .05 alpha: No

Day 10: Advanced Designs

Testing for interactions

BRIEFLY

Complex models build on the principles we already discussed. Although their calculation is beyond the scope of this discussion (that's what computers are for), here is an introduction to procedures that use multiple predictors, multiple criteria and multivariate techniques to test interactions between model components.

There are four types of complex models I'd like to review:

Factorial ANOVA
Multiple regression
Multivariate analysis
Causal modeling

INTRODUCTION

Until now, our models have been quite simple. One individual, one group, or one variable predicting another. We have explored the levels of measurement, the importance of theories and how to convert theoretical constructs into model variables. We have taken a single variable, plotted its frequency distribution and described its central tendency and dispersion. We have used percentiles and z-scores to describe the location of an individual score in relation to the group.

In addition to single variable models, we studied two variable models, such as correlations, regressions, t-tests and one-way ANOVAs. We have laid a thorough foundation of research methods, experimental design, and descriptive and inferential statistics.

Despite their simplicity, these procedures are very useful. You can use a correlation to measure the reliability and validity of a test, machine or system of management, training or production. You can use a linear regression to time date a rare archaeological find, predict the winner of a race or analyze a trend in the stock market. You can use the t-test to test a new drug against a placebo or compare 2 training conditions. You can use the 1-way ANOVA to test several psychotherapies, compare levels of a drug or brands of computers.

Also, the procedures you've studied so far can be combined into more complex models. The most complex models have more variables but they are variations of the themes you've already encountered.

A factorial AVOVA is like combining 1-way ANOVAs together. The purpose of combining the designs is to test for interactions. A 1-way ANOVA can test to see if different levels of salt will influence customer preference, but what happens if the soft drink is both salty and sweet?

Factorial designs

A factorial ANOVA tests the impact of 2 or more independent variables on one dependent variable. It tests the influence of many discrete variables on one continuous variable. It has multiple independent variables and one dependent variable.

A 1-way ANOVA model tests multiple levels of 1 independent variable. Let's assume the question is does stress cause people to perform multiplication problems. Subjects are randomly assigned to a treatment level (high, medium and low, for example) of one independent variable (stress, for example). And their performance on one dependent variable (number of errors) is measured. If stress impacts performance, you would expect errors to increase with the level of stress. The variation between the cells is due to the treatment given. Variation within each cell is thought to be due to random chance.

A 2-way ANOVA has 2 independent variables. Here is a design which could look at gender (male; female) and stress (low, medium and high): It is called a 2 x 3 ("two by three") factorial design. If each cell contained 10 subjects, there would be 60 subjects in the design.

A design for the amount of student debt (low, medium and high) and year in college (frosh, soph, junior and senior) would have 1 independent variable (debt) with 3 levels and 1 independent (year in school) with 4 levels. This is a 3x4 factorial design.

Notice that the number (3, 4, etc.) tells how many levels in the independent variable. The number of numbers tells you how many independent variables there are. A 2x4 has 2 independent variable. A 3x7 has 2 independent variables (one with 3 levels and one with 7 levels). A 2x3x4 factorial design has 3 independent variables.

Factorial designs can do something 1-way ANOVAs can't. Factorial designs can test the interaction between independent variables. Taking pills can be dangerous and driving can be dangerous; but it often is the interaction between variables that interests us the most.

Analyzing a 3x4 factorial design involves 3 steps: columns, rows and cells. The factorial ANOVA treats the columns of the design as if each column was a different group. Like a 1-way ANOVA, this main effect tests the columns as if the rows didn't exist.

The second main effect (rows) is tested as if each row was a different group. It tests the rows as if the columns didn't exist. Notice that each main effect is like doing a separate 1-way ANOVA on that variable.

The cells also are tested to see if one cell is significantly larger (or smaller) than the others. This is a test of the interaction and checks to see if a single cell is significantly different from the rest. If one cell is significantly higher or lower than the rest, then it is the result of a combination of the independent variables.

Multiple Regression

An extension of simple linear regression, multiple regression is based on observed data. In the case of multiple regression, two or more predictors are used; there are multiple predictors and a single criterion.

Let's assume that you have selected 3 continuous variables as predictors and 1 continuous variable as criterion. You might want to know if gender, stress and time of day impact typing performance.

Each predictor is tested against the criterion separately. If a single predictor appears to be primarily responsible for changes in the criterion, its influence is measured. Every combination of predictors is also tested. So both main effects and interactions can be tested. If this sounds like a factorial ANOVA, you're absolutely correct.

You could think of multiple regression and ANOVA as siblings. factorial ANOVAs use discrete variables; multiple regression uses continuous variables. If you were interested in using income as one of your predictors (independent variables), you could use discrete categories of income (high, medium and low) and test for significance with an ANOVA. If you wanted to use measure income on a continuous variable (actual income earned), the procedure would be a multiple regression.

You also could think of multiple regression and the parent of ANOVA. Analysis of Variance is actually a specific example of multiple regression; it is the discrete variable version. Analysis of Variance uses categorical predictors. Multiple regression can use continuous or discrete predictors (in any combination); it is not restricted to discrete predictors.

Both factorial ANOVA and multiple regression produce a F statistic and both have only one outcome measure. Both produce a F score that is compared to the Critical Values of F table. Significance is ascribed if the calculated value is larger than the standard given in the table.

Both procedures have only one outcome measure. There may be many predictors in a study but there is only one criterion. You may select horse weight, jockey height, track condition, past winning and phase of the moon as predictors of a horse race but only one outcome measure is used. Factorial ANOVA and multiple regression are multiple predictor-single criterion procedures.

Multivariate Analysis

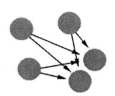

Sometime called MANOVA (pronounced *man-o-va*), multivariate analysis is actually an extension of multiple regression. Like multiple regression, multivariate analysis has multiple predictors. In addition to multiple predictors, multivariate analysis allows multiple outcome measures.

Now it is possible to use gender, income and education as predictors of happiness AND health. You are no longer restricted to only a single criterion. With multivariate analysis, the effects and interactions of multiple predictors can be examined. And their impact on multiple outcomes can be assessed.

The analysis of a complex multiple-predictor multiple-criteria model is best left to a computer but the underlying process is the calculation of correlations and linear regressions. As variables are selected for the model, a decision is made whether it is a predictor or a criterion. Obviously, aside from the experimenter's theory, the choice of predictor or criterion is arbitrary. In multivariate analysis, a variable such as annual income could be either a predictor or a criterion.

Complex Modeling

There are a number of statistical procedures at the high end of modeling. Relax! You don't have to calculate them. I just want you to know about them.

In particular, I want to make the point that there is nothing scary about the complex models. They are involved and require lots of tedious calculations but that's why God gave us computers. Since we are blessed to have stupid but remarkably fast mechanical slaves, we should let them do the number crunching.

It is enough for us to know that a complex model—at its heart—is a big bundle of correlations and regressions. Complex models hypothesize directional and nondirectional relationships between variables. Each factor may be measured by multiple measures. Intelligence might be defined as the combination of 3 different intelligence tests, for example.

Income might be a combination of both salary plus benefits minus vacation. And education might be years in school, number of books read and number of library books checked out. The model, then, becomes the interaction of factors that are more abstract than single variable measures.

Underlying the process, however, are principles and procedures you already know. Complex models might try to determine if one more predictor helps or hurts the model but they are evaluated just like correlations: percentage of variance accounted for by the relationships.

UNDERSTAND

The story is in the details. And different statistical procedures show different amounts of detail. So choose the statistical procedure that shows the amount of detail you want to see.

Illustration 1: A visual image should not be sketchy.:

It should have enough detail to convey meaning:

Even if it doesn't look like reality:

Illustration 2: A case study should not be sketchy. It should describe the behavior and symptoms of the client and provide a thorough family history. It should be more like a book than a postcard.

REMEMBER

Basic Facts

Review the Basic Facts test. You should now be able to recall and understand everything on it.

Formulas

Review the formulas. You don't have to be able to recall the formulas from memory but you should now be able to choose the correct formula for each procedure.

Terms

Review the terms. You don't have to be able to list them from memory but you should now be able to define and distinguish between each of them.

DO

Take the practice final at the end of this chapter. Don't memorize the wording of the questions but focus on the process. You should now be able to choose the proper procedure for a given situation, be able to accurately calculate the statistics and be able to interpret the results.

SUMMARY

We began the course looking at a single variable. You discovered its central tendency (mean, median and mode) and its dispersion (range, MAD, SS, variance and standard deviation). You learned how to compare your score to the group by using percentiles and areas under the curve. With a z-score, you learned how many standard deviations a score is from its mean. You encountered the normal curve, bimodal distributions, and positively- and negatively-skewed distributions.

With correlation and regression, the single variable model was expanded to 2 dependent variables. You observed the pattern of a scatterplot, learned to estimate the strength of a correlation, and made predictions (extrapolations and interpolations) based on their pattern of relationship. You also tested the significance of the 2-DV model with an Analysis of Regression.

The t-test was also a 2-variable model: 1 independent variable and 1 dependent variable. The independent variable was split into 2 conditions (treatment and control) and subjects were randomly assigned to each. Building on the t-test, you learned to extend the 2-variable model by splitting the IV into more parts for a 1-Way ANOVA.

In advanced models, the basic principles remain the same but more variables are added. In a 2-Way ANOVA, 2 independent variables are used to predict the performance of 1 dependent variable. Factorial ANOVAs get more complex (adding more and more IVs) but each has only 1 dependent variable.

Multiple regression uses continuous variables as predictors and mirrors the designs of factorial ANOVAs. In fact, factorial ANOVAs are simply a subset of multiple regression (MR) procedures. ANOVA uses discrete (high, medium, low) predictors and MR can use discrete or continuous predicators (or any combination of them).

In multivariate analysis (often called MANOVA), multiple predictors (like those in factorial ANOVA and MR) are used to predict multiple outcomes.

Causal modeling (and all the other super-complex designs) use several measures of each factor and then the factors are treated like predictors and criteria. They are giant MANOVAs, where each factor is composed of 2 or 3 (or more) different measures.

Practice Final Exam

1. Which of the following shows the interaction of 2 independent variables:
   a. t-test
   b. one-way ANOVA
   c. factorial ANOVA
   d. all of the above

2. In a 2x3x4 factorial design, how many independent variables are there:
   a. 1
   b. 3
   c. 9
   d. 24

3. How many dependent variables are in a 3x3 factorial design:
   a. 1
   b. 2
   c. 3
   d. 6

4. In a factorial ANOVA, degrees of freedom for SS_{total} equals:
   a. N
   b. N minus 1
   c. k minus 1
   d. N minus k

5. Which would use multiple measures of a factor (such as different intelligence tests) as a criterion:
   a. causal modeling
   b. 1-way ANOVA
   c. Analysis of Regression
   d. 2-way ANOVA

6. The type of relationships between model components is determined by our:
   a. theoretical questions
   b. empirical analysis
   c. sampling error
   d. statistical bias

7. In an independent measures t-test with n = 10 and N=20, how many degrees of freedom are in the study:
   a. 10
   b. 18
   c. 19
   d. 20

8. When a frequency distribution is positively skewed, the mean is:
   a. higher than the median
   b. lower than the median
   c. same as the median
   d. same as the mode

9. If a manager randomly assigns phone calls to all 10 customer service representatives, which of the following should be used to test for significant differences in the speed of call handling:
   a. multiple regression
   b. regression
   c. t-test
   d. 1-way ANOVA

10. Which of the following represents the height of a frequency distribution:
    a. mean
    b. median
    c. mode
    d. standard deviation

11. If a 1-way ANOVA is performed on 4 personality types, how many degrees of freedom are associated with the Between Sum of Squares:
    a. 1
    b. 3
    c. 4
    d. can't tell without knowing N

12. In a 2x3 factorial design (Gender and Wing Span of butterflies), a significant interaction would indicate that there is a significant difference between:
    a. male and female butterflies
    b. small- and large-winged butterflies
    c. medium- and large-winged butterflies
    d. a single cell and the rest of the cells

13. Which is most affected by outlying scores:
    a. mean
    b. median
    c. mode
    d. A, B and C are equally affected

14. Rejecting the null when you should have accepted it is:
    a. Type I error
    b. Type II error
    c. Type III error
    d. Typo error

15. How many dependent variables are in a 2x 3x3 factorial design:
    a. 1
    b. 3
    c. 6
    d. 9

16. A test which described a restricted range of people will probably result in a distribution which is:
    a. linear
    b. curvilinear
    c. skewed
    d. normal

17. The more samples taken, the more normal the curve looks, according to the:
    a. critical-limit hypothesis
    b. maximum-limit theorem
    c. unbiased-limit theorem
    d. central-limit theorem

18. Which correlation coefficient shows the greatest amount of relationship:
    a. .45
    b. .56
    c. .71
    d. -.89

19. A nondirectional test of significance is said to be:
    a. no-tailed
    b. one-tailed
    c. two-tailed
    d. three-tailed

20. Which of the following is defined as a cumulative frequency divided by N (times 100):
    a. frequency distribution
    b. percentile
    c. percent
    d. proportion

21. From one standard deviation above the mean to one standard deviation below the mean accounts for what percent of scores:
    a. 17%
    b. 25%
    c. 68%
    d. 95%

22. A correlation between a discrete and a continuous variable is called:
    a. phi
    b. Pearson r
    c. point biserial
    d. confidence level

23. If 4 car colors are tested to see which lasts the longest in the sun, which of the following should be used:
    a. correlation
    b. t-test
    c. 1-way ANOVA
    d. factorial AVOVA

24. This distribution is best described as:

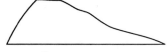

   a. normal
   b. positively skewed
   c. negatively skewed
   d. strangely familiar

25. A t-test compares:
   a. two medians
   b. two modes
   c. two means
   d. two standard deviations

26. To compare an individual to a group, use a:
   a. t-test
   b. z-score
   c. correlation
   d. regression

27. Scores projected between data points are:
   a. interpolated
   b. extrapolated
   c. innovated
   d. insubstantiated

28. In an actual study, annual income could be:
   a. a predictor
   b. a criterion
   c. an intervening variable
   d. all of the above

29. In the simplest case, the probability of A and B occurring is calculated by:
   a. adding the probabilities
   b. subtracting the probabilities
   c. multiplying the probabilities
   d. dividing the probabilities

30. When the null hypothesis is true, the expected value for an independent measures t statistic is:
   a. 0
   b. +1.0
   c. -1.0
   d. +1 or -1

31. How many df to test a Pearson r :
   a. N
   b. N-1
   c. N-2
   d. N-k

32. Which could be a point-biserial correlation:
   a. -1.2
   b. -.87
   c. -3.0
   d. 4.27

33. Which shows a correlation's strength:
   a. magnitude
   b. stanine
   c. stature
   d. sign

34. Which is the sum of the squared deviations:
   a. mean
   b. mean variance
   c. sum of squares
   d. standard deviation

35. Which should be used to measure reliability:
   a. multiple regression
   b. regression
   c. correlation
   d. t-test

36. A variable whose levels are described as "high," "medium" and "low" is:
   a. reliable
   b. discrete
   c. continuous
   d. logarithmic

37. A z-score of -1.5 on a test with a mean of 100 and a standard deviation of 10 would equal a score of:
   a. 85
   b. 100
   c. 101.5
   d. 115

- 220 -

38. T-tests are calculated like:
   a. point estimations
   b. sum of squares
   c. degrees of freedom
   d. z-scores

39. Which of the following is an example of a grouped frequency distribution:
   a. data matrix
   b. correlation
   c. regression
   d. histogram

40. Which do we select, manipulate or induce:
   a. independent variable
   b. dependent variable
   c. moderator variable
   d. suppressor variable

41. Which should be used to make predictions:
   a. correlation
   b. regression
   c. correlated t-test
   d. independent t-test

42. Which is the coefficient of determination:
   a. t
   b. r
   c. r^2
   d. $1-r^2$

43. The number on your race car is:
   a. nominal
   b. ordinal
   c. interval
   d. ratio

44. A repeated measures design reduces error variability due to:
   a. mean of D
   b. degrees of freedom
   c. the effect of the treatment
   d. individual differences

45. To test the interaction of diet and exercise on the maze running behavior of rats, a researcher could use a:
   a. t-test
   b. correlation
   c. 1-Way ANOVA
   d. 2x3 ANOVA

46. Which shows the interaction of 2 independent variables:
   a. regression
   b. 1-Way ANOVA
   c. t-test
   d. factorial ANOVA

47. To compare 2 dependent variables a researcher would use a:
   a. t-test
   b. z-score
   c. correlation
   d. 1-way ANOVA

48. To predict future success in psychology from current GPA, a researcher would use a:
   a. t-test
   b. correlation
   c. regression
   d. 1-way ANOVA

49. Which of the following **cannot** be an appropriate Sum of Squares of Memory:
   a. 21.5
   b. 1.27
   c. 129.2
   d. -42

50. What percentage of scores are beyond a z of +1.
   a. 14%
   b. 16%
   c. 34%
   d. 48%

51. As a pigeon seller, you are interested in how weight affects sale price. Here is a sample of information from the last few months:

Weight	Price
2	12
4	10
7	8
9	10
13	5

What is the SS for Weight: _____

What is the variance of Weight: _____

What is the range of Weight: _____

What is the SS for Price: _____

What is the SSxy: _____

Since you are interested in how well one rating acts as a linear predictor of another, which of the following tests should you perform:
  a.   t-test
  b.   ANOVA
  c.   correlation
  d.   regression
  e.   multiple regression

Perform the comparison you selected in the item above. Select only the appropriate one(s). What was the result of your calculations?

   $a =$

   $b =$

   $r =$

   $t =$

   $F =$

If weight equals 3, what would you predict price to be?

52. As a biologist, you are interested in the ability of pigeons to run and fly. You have measured their performance on each task, and now hope to find how related these two variables are. The numbers below represent the number of feet per activity.

Walk	Fly
15	11
8	9
4	7
11	10
2	3

What is the sum of Walk:_____

What is the mean of Walk: _____

What is the SS for Walk:_____

Since you are interested in commonality, which of the following tests should you perform:
      a. multiple regression
      b. regression
      c. correlation
      d. t-test
      e. ANOVA

Perform the comparison you selected in the item above. What was the result of your calculation (select only the appropriate ones):

      a =

      b =

      r =

      t =

      F =

How many degrees of freedom are in this study?

What is the critical value for this statistic?

Is there a significant relationship between these variables at the .05 alpha level?

What percentage of variance is shared by the two variables?

Calculate the coefficient of non-determination: _____

53. As a chef, you want to find the best way to serve pigeons. You catch twelve of them and randomly assign them to fried and steamed. That evening you serve both in your cafe for 6 hours; which version sells best?

Fried	Steamed
5	8
6	4
13	6
7	3
9	4

What is the median for Fried: _____

What is the mean for Fried: _____

What is the SS for Fried: _____

What is the median for Steamed:_____

What is the mean for Steamed: _____

Since you are interested in which group did significantly better, which of the following tests should you perform:

      a.t-test
      b.ANOVA
      c.correlation
      d.regression
      e.multiple regression

Perform the comparison you selected in the item above. Select only the appropriate one(s). What was the result of your calculation?

      a =

      b =

      r =

      t =

      F =

What is the critical value?

Is there a significant difference between the two groups?

54. As a regional manager, you're interested in finding the fastest way to deliver sales info to the home office. After randomly assigning messages to 3 methods of communication, you measure the number of hours it takes for the data to be delivered. Here's what you found:

Walking	Pigeon	Email
13	9	5
13	5	7
15	5	4
11	1	3
9	2	2
5	8	3

Since you are interested in which method did best, which of the following tests should you perform:

    a.t-test
    b.ANOVA
    c.correlation
    d.regression
    e.multiple regression

How many independent variables are in this design: _____

What is the dependent variable in this design: _____

Based on above data, complete the following summary table:

	SS	df	ms
Between	____	____	____
Within	____	____	____
TOTAL	____	____	

What is the F for this test:

What is the critical value:

Is the F significant:

What should be done next:

55. You wonder if the amount of peanut butter has an significant impact on the rating of your new cookie's popularity. Use the following data to complete the summary table:

r= .55
SSx = 60
SSy = 50
N = 14

	SS	df	ms
Regression	____	____	____
Error	____	____	____
TOTAL	____	____	

What is the predictor in this study:

What is the criterion in this study:

What is the F for this test: _____

What is the critical value for F: _____

Is F significant at .05 alpha: _____

What is your conclusion:

Answers

Practice Final Exam
1. c, 2. b, 3. a, 4. b, 5. a, 6. a, 7. b, 8. a, 9. d, 10. c
11. b, 12. d, 13. a, 14. a, 15. a, 16. c, 17. d, 18. d, 19. c, 20. b
21. c, 22. c, 23. c, 24. b, 25. c, 26. b, 27. a, 28. d, 29. c, 30. a
31. c, 32. b, 33. a, 34. c, 35. c, 36. b, 37. a, 38. d, 39. d, 40. a
41. b, 42. c, 43. a, 44. d, 45. d, 46. d, 47. c, 48. c, 49. d, 50. b

51. Regression

What is the SS for Weight:	74
What is the variance of Weight :	18.50
What is the range of Weight:	11
What is the SS for Price:	28
What is the SSxy:	- 40

a = 12.78
b = - .54
If weight equals 3, what would you predict price to be? 11.16

52. Correlation

What is the sum of Walk:	40
What is the mean of Walk:	8
What is the SS for Walk:	110

r = .92
How many degrees of freedom in this study? 8
What is the critical value for this statistic?   .63
Is there a significant relationship between these variables at the .05 alpha level? Yes
What percentage of variance is shared by the two variables? .85
Calculate the coefficient of non-determination: .15

53. t-test

What is the median for Fried:	7
What is the mean for Fried:	8
What is the SS for Fried:	40
What is the median for Steamed:	4
What is the mean for Steamed:	5
What is the critical value:	2.31

t = 1.79
Is there a significant difference between the two groups? No

54. 1-Way Analysis of Variance

How many independent variables are in this design: 1 (communication)

What is the dependent variable in this design: hours

	SS	df	ms
Between	172	2	86
Within	130	15	8.67
TOTAL	302	17	

What is the F for this test: 9.92

What is the critical value: 3.68

Is the F significant? Yes

What should be done next: t-tests

55. Analysis of Regression

	SS	df	ms
Between	15.13	1	15.13
Within	34.88	12	2.91
TOTAL	50	13	

What is the predictor in this study: Peanut butter

What is the criterion in this study: Popularity

What is the F for this test: 5.20

What is the critical value for F: 4.75

Is F significant at .05 alpha: Yes

What is your conclusion:

The amount of peanut butter has a significant impact on the popularity of your cookies. The data fits the model of a straight line.

Basic Facts Test (50 things you should keep in your head)

1. List three measures of central tendency:
   a.
   b.
   c.

2. List five measures of dispersion:
   a.
   b.
   c.
   d.
   e.

3. Complete the following:
   a. Theories are composed of:
   b. Models are composed of:
   c. Laws:
   d. Principles:
   e. Beliefs:

4. List six criteria for evaluating theories:
   a.
   b.
   c.
   d.
   e.
   f.

5. List four levels of measurement:
   a.
   b.
   c.
   d.

6. List three types of correlation AND the kind of variables with which they are used:
   a.
   b.
   c.

7. List two characteristics of a data matrix:
   a.
   b.

8 List six things associated with a linear regression:
- a.
- b.
- c.
- d.
- e.
- f.

9. List four types of variables:
- a.
- b.
- c.
- d.

10. List and describe nine applications of the General Linear Model to continuous and discrete variables:

Continuous Models compare:
- a. causal modeling
- b. multivariate analysis
- c. multiple regression
- d. regression
- e. correlation
- f. frequency distribution

Discrete Models Compare:
- a. t-test
- b. one-way ANOVA
- c. factorial ANOVA

Basic Facts Test (50 things you should keep in your head)
ANSWERS

1. List three measures of central tendency:
   a. Mean
   b. Median
   c. Mode

2. List five measures of dispersion:
   a. Sum of Squares
   b. Variance
   c. Standard Deviation
   d. Mean Variance
   e. Range

3. Complete the following:
   a. Theories are composed of: Constructs
   b. Models are composed of: Variables
   c. Laws: Accuracy beyond doubt
   d. Principles: Some predictability
   e. Beliefs: Personal opinion

4. List six criteria for evaluating theories:
   a. Clear
   b. Useful
   c. Summarize facts
   d. Small number of assumptions
   e. Internally consistent
   f. Testable hypotheses

5. List four levels of measurement:
   a. Nominal
   b. Ordinal
   c. Interval
   d. Ratio

6. List three types of correlation and the kind of variables with which they are used:
   a. phi   2 discrete variables
   b. Pearson r   2 continuous variables
   c. point-biserial   1 continuous and 1 discrete variable

7. List two characteristics of a data matrix:
   a. Columns are attributes
   b. Rows are entities (subjects)

8 List six things associated with a linear regression:
    a. Intercept
    b. Slope
    c. Interpolate
    d. Extrapolate
    e. Least squares criterion
    f. Standard error of estimate

9. List four types of variables:
    a. Independent
    b. Dependent
    c. Modifier
    d. Intervening

10. List and describe nine applications of the General Linear Model to continuous and discrete variables:

Continuous Models compare:
    a. causal modeling   Multiple measures of a factor
    b. multivariate analysis   Multiple predictors; multiple criteria
    c. multiple regression   Multiple predictors; single criterion
    d. regression   Single predictor; single criterion
    e. correlation   Two regressions
    f. frequency distribution One variable (predictor or criterion)

Discrete Models Compare:
    a. t-test   2 means; 1 independent variable
    b. one-way ANOVA   3 or more means; 1 independent variable
    c. factorial ANOVA   2+ means on 2+ independent variables

TABLES
TABLE A: AREAS UNDER THE NORMAL CURVE

	zBetween	Beyond	Percentile		zBetween	Beyond	Percentile
0	0.0000	0.5000	0.5000	1.10	0.3643	0.1357	0.8643
0.01	0.0040	0.4960	0.5040	1.15	0.3749	0.1251	0.8749
0.02	0.0080	0.4920	0.5080	1.20	0.3849	0.1151	0.8849
0.03	0.0120	0.4880	0.5120	1.25	0.3944	0.1056	0.8944
0.04	0.0160	0.4840	0.5160	1.30	0.4032	0.0968	0.9032
0.05	0.0199	0.4801	0.5199	1.35	0.4115	0.0885	0.9115
0.06	0.0239	0.4761	0.5239	1.40	0.4192	0.0808	0.9192
0.07	0.0279	0.4721	0.5279	1.45	0.4265	0.0735	0.9265
0.08	0.0319	0.4681	0.5319	1.50	0.4332	0.0668	0.9332
0.09	0.0359	0.4641	0.5359	1.55	0.4394	0.0606	0.9394
0.10	0.0398	0.4602	0.5398	1.60	0.4452	0.0548	0.9452
0.15	0.0596	0.4404	0.5596	1.65	0.4505	0.0495	0.9505
0.20	0.0793	0.4207	0.5793	1.70	0.4554	0.0446	0.9554
0.25	0.0987	0.4013	0.5987	1.75	0.4599	0.0401	0.9599
0.30	0.1179	0.3821	0.6179	1.80	0.4641	0.0359	0.9641
0.35	0.1368	0.3632	0.6368	1.85	0.4678	0.0322	0.9678
0.40	0.1554	0.3446	0.6554	1.90	0.4713	0.0287	0.9713
0.45	0.1736	0.3264	0.6736	1.95	0.4744	0.0256	0.9744
0.50	0.1915	0.3085	0.6915	1.96	0.4750	0.0250	0.9750
0.55	0.2088	0.2912	0.7088	2.00	0.4772	0.0228	0.9772
0.60	0.2257	0.2743	0.7257	2.10	0.4821	0.0179	0.9821
0.65	0.2422	0.2578	0.7422	2.20	0.4861	0.0139	0.9861
0.70	0.2580	0.2420	0.7580	2.30	0.4893	0.0107	0.9893
0.75	0.2734	0.2266	0.7734	2.40	0.4918	0.0082	0.9918
0.80	0.2881	0.2119	0.7881	2.50	0.4938	0.0062	0.9938
0.85	0.3023	0.1977	0.8023	2.60	0.4953	0.0047	0.9953
0.90	0.3159	0.1841	0.8159	2.70	0.4965	0.0035	0.9965
0.95	0.3289	0.1711	0.8289	2.80	0.4974	0.0026	0.9974
1.00	0.3413	0.1587	0.8413	2.90	0.4981	0.0019	0.9981
1.05	0.3531	0.1469	0.8531	3.00	0.4987	0.0013	0.9987

TABLE B:
Critical Values of the Pearson r
2-tailed tests of r

df	.05 alpha
1	0.997
2	0.950
3	0.878
4	0.811
5	0.755
6	0.707
7	0.666
8	0.632
9	0.602
10	0.576
11	0.553
12	0.532
13	0.514
14	0.497
15	0.482
16	0.468
17	0.456
18	0.444
19	0.433
20	0.423
25	0.381
30	0.349
35	0.325
40	0.304
45	0.288
50	0.273
60	0.250
70	0.232
80	0.217
90	0.205
100	0.195

TABLE C:
Critical Values of Student's t
2-tailed t-tests

df	.05 alpha
1	12.71
2	4.30
3	3.18
4	2.78
5	2.57
6	2.45
7	2.37
8	2.31
9	2.26
10	2.23
11	2.20
12	2.18
13	2.16
14	2.15
15	2.13
16	2.12
17	2.11
18	2.10
19	2.09
20	2.09
21	2.08
22	2.07
23	2.07
24	2.06
25	2.06
26	2.06
27	2.05
28	2.05
29	2.05
30	2.04
40	2.02
60	2.00
120	1.98
infinity	1.96

Adapted from the Critical Values of the Pearson r, Table VII from Fisher & Yeates' Statistical Tables for Biological, Agricultural and Medical Research, Longman Group Ltd, London, 1974. (Pearson Education Ltd.). Used by permission.

Adapted from "Critical Values of the Student t" which appeared in E.T. Federighi's 1959 article "Extended tables of the percentage points of Student's t distribution." in the Journal of the American Statistical Association, 54, 683-688. Used by permission.

TABLE D:
Critical Values of F
.05 alpha

Degrees of Freedom Regression
Degrees of Freedom Between

df	1	2	3	4	5
1	161	200	216	225	230
2	18.51	19.00	19.16	19.25	19.3
3	10.13	9.55	9.28	9.12	9.01
4	7.71	6.94	6.59	6.39	6.26
5	6.61	5.79	5.41	5.19	5.05
6	5.99	5.14	4.76	4.53	4.39
7	5.59	4.47	4.35	4.12	3.97
8	5.32	4.46	4.07	3.84	3.69
9	5.12	4.26	3.86	3.63	3.48
10	4.96	4.10	3.71	3.48	3.33
11	4.84	3.98	3.59	3.36	3.20
12	4.75	3.88	3.49	3.26	3.11
13	4.67	3.80	3.41	3.18	3.02
14	4.60	3.74	3.34	3.11	2.96
15	4.54	3.68	3.29	3.06	2.90
16	4.49	3.63	3.24	3.01	2.85
17	4.45	3.59	3.20	2.96	2.81
18	4.41	3.55	3.16	2.93	2.77
19	4.38	3.52	3.13	2.90	2.74
20	4.35	3.49	3.12.	2.87	2.71

Degrees of Freedom Error
Degrees of Freedom Within

Adapted from "Critical Values of the F distribution," SPSS.com. Used by permission.

Formulas

DESCRIPTIVE

Sum of Squares

$$SS = \sum X^2 - \frac{\left(\sum X\right)^2}{N}$$

variance = SS/df
   population
   sample

standard deviation
   s (population)
   s (sample)

INFERENTIAL

$$z = \frac{X - \overline{X}}{s}$$

$$t = \frac{\overline{X}_1 - \overline{X}_2}{\sqrt{\dfrac{SS_{x_1} + SS_{x_2}}{n(n-1)}}}$$

$$F = \frac{\text{Mean Squares}_{(Between)}}{\text{Mean Square}_{(Within)}}$$

Mean Square = SS/df

SSw = SS1 + SS2 + SS3+...

$$SS_b = \left(\frac{\left(\sum X_1\right)^2 + \left(\sum X_2\right)^2 + \left(\sum X_3\right)^2 + ...}{n}\right) - \frac{\left(\sum X\right)^2}{N}$$

$$SS_t = \sum X^2_1 + \sum X^2_2 + \sum X^2_3 + ... - \frac{\left(\sum X\right)^2}{N}$$

CORRELATION

$$r = \frac{SSxy}{\sqrt{SSxSSy}}$$

$$r = \frac{\sum XY - \frac{(\sum X)(\sum Y)}{N}}{\sqrt{(\sum X^2 - \frac{(\sum X)^2}{N})(\sum Y^2 - \frac{(\sum Y)^2}{N})}}$$

REGRESSION

Formula for a line:

$$Y' = a + bX$$

a = intercept

$$a = \overline{Y} - b\overline{X}$$

b = regression coefficient; slope

$$b = \frac{SS_{xy}}{SS_x}$$

Standard error of estimate

Feedback

You may contact the author by email. Put "Safari" in the subject line and send your thoughts, suggestions and comments to feedback@kentangen.com. Also, feel free to browse his website at www.kentangen.com. For additional explanations, practice problems simulations and exercises, go to www.statisticssafari.com.

NOTES

If you have trouble following formulas, use this space to jot reminders to youself on the specific steps in each procedure. Sometimes it helps to put things in your own words.

ISBN: 0-9765360-2-1
Printed in the United States of America

Printed in the United States
47116LVS00001BA/285-294